AMERICAN AVIATION ICONS 2

GRUMMAN

F9F PANTHER & COUGAR

THEODORE GIANNA

THUNDERBOLT 47 PUBLICATIONS

Copyright © 2022 Theodore Gianna. (THUNDERBOLT 47 PUBLICATIONS)
This publication is produced under the jurisdiction of international copyright laws. All rights reserved. No part of this publication may be reproduced, distributed, or transmitted in any form or by any means, including photocopying, recording, electronic, mechanical methods, or otherwise, without the express permission of the publishers and copyright holders.
 Every effort has been made to make this book's information as complete and accurate as possible. However, the Publisher and Author accept no liability for any errors nor be held responsible for any subsequent use made of the published information, data, or specific details within this publication.

ISBN: 978-0-6489379-8-2 (Paperback)
First printing edition 2022.

CONTRIBUTORS

I would like to thank the National Naval Aviation Museum, especially Jared Galloway and Bob Adams, for providing the many fascinating photos in the book. I would also like to thank the Naval History and Heritage Command for the historical naval aircraft images.

INTRODUCTION

This work is the second book in a series of American military aircraft from the second world war to the end of the third generation of American military jets, which ended about 1975.

This book offers a concise history of two exceptional Grumman models, the first of which got the first 'jet' kill for the US Navy in combat during the Korean War - the 'F9F Panther'. It was an extremely versatile jet that outperformed the older F-80 and was as good as, if not better than, the Republic Thunderjet. Both of which also fought in the Korean conflict.
   The Panther's success prompted Grumman to create a variant with swept wings and pitch it to the Navy. This second model was accepted enthusiastically by the Navy – the 'Cougar.' It was quicker and more streamlined than the Panther, although it was not involved in a major war. However, there was one exception; some TF-9J Cougars, which were the trainer version, were used in small numbers during the Vietnam War as Forward Air Control (FAC) aircraft. Nevertheless, the Panther and the Cougar performed excellently in service with the US Navy. Many were surprised by the strength and resilience of the two jets, but not the Navy personnel. "It's Grumman engineering." was their reply. The Grumman company that produced the jets had a record of manufacturing robust, rugged, and reliable aircraft. I totally agree; the Panther and Cougar jets were Grumman aircraft, through and through.

GRUMMAN F9F PANTHER

Grumman Aircraft Engineering Corporation was formed in 1929 after the merger of the Loening and Keystone companies and quickly became one of America's most successful aircraft manufacturers. Operations started on January 2, 1930.

Loening, who had previously been active in producing amphibian aircraft, created the company's first successful aircraft, the Grumman Model A amphibian float gear. The Grumman Model A started a unique relationship between Grumman and the US Navy. Grumman submitted its aircraft to the Navy and caused sufficient interest to gain Grumman a contract for two units. The Navy soon increased its order to eight aircraft, and from that period, the US Navy and Grumman began to have a close relationship that continued for many decades. In addition, the Navy saw Grumman as a dependable company with a knack for manufacturing reliable, sturdy, and safe aircraft.

During the Second World War, Grumman produced two of the finest carrier-based fighter aircraft; the 'F4F' Wildcat,' the F6F Hellcat,' and the excellent torpedo bomber, the 'TBF Avenger.' It also manufactured the 'F8F' Bearcat.' Unfortunately, the latter did not see any action in World War II because it was put into service just after the War ended. It was also Grumman's last piston-engine fighter. Moreover, after serving only a few years in Navy and Marine squadrons, the 'Bearcat' had to make way for the first of the naval jet planes with their novel engines, the most brilliant of which was its direct replacement, the F9F Panther.

The qualitative leap forward required by this new method of propulsion that the F9F possessed affected not only the planes themselves but also the aircraft carriers that transported them, which had to come to grips with the latest technology during an initial adjustment period. The installation of new steam catapults for jets on the carriers was essential because of the early jet engines' low acceleration and takeoff speeds. Furthermore, the setting up of enhanced deceleration devices was made necessary by the high landing speeds of the jets. By the time the first Panthers made their appearance, these two elements had already been put in place.

XF9F-2

As early as 1946, the Grumman engineers started work on a new project. The new design was based on the F8F' Bearcat.' The Bearcat was only a single-engine, single-seat fighter. Its successor, however, was to be a side-by-side, two-seat, carrier-based, high-performance night fighter powered by four Westinghouse 19X B-2B (Navy: J-30) axial-flow turbojets mounted beneath the 55-foot 6-inch wing. The new project was given the designation XF9F-1, but it never saw the light of day, as it was canceled before the design work was completed.

Due to the availability of the more powerful British Rolls Royce Nene engine with its 5,000 lb. thrust, Grumman decided to create a new single-engine day fighter based around that engine, Model G-79. It was to become the F9F Panther. Subsequently, the new design was proposed to the Navy, and the Navy Bureau of Aeronautics approved it wholeheartedly. As a result, three prototypes were ordered: 122475, 122476, and 122477. The first and last prototypes were converted into XF9F-2s, while the second prototype was the only XF9F-3. Later, the two XF9F-2 prototypes were powered by the imported British Rolls-Royce Nene jet engines. Finally, the Allison J33-A-8, comparable in size but less powerful than the Nene, was chosen for the

This is the prototype XF9F-2, flown by Grumman's test pilot Corwin Henry Meyer. The first two prototypes (BuNo 122475, 122477) were powered by Rolls-Royce Nene engines and first flew on November 21 and 26, 1947. Note that this XF9F-2 is without the wingtip tanks which became standard in all production models. (NAID 176250458)

XF9F-3 prototype and production model F9F-3. That meant the only difference between the F9F-2 and F9F-3 models was the installation of different engines.

Additionally, from the beginning of the F9F program, the most visible improvement on the Panthers was the addition of the permanently fixed, wingtip-mounted external fuel tanks.

F9F-2/F-3

The XF9F-2 design was still being studied informally in June 1946, although some conclusions had been reached. By August, the design had been formalized, and two Grumman engineers traveled to Britain to evaluate the Nene engine.

On October 2, 1946, the Navy's Bureau of Aeronautics (BuAer) made two crucial decisions. First, it recommended the designation XF9F-1 should be replaced with XF9F-2. That was done to confirm the abandonment of the four-engine night fighter in favor of the single-engine day fighter. The second decision was to change the engines. The Bureau of Aeronautics favored substituting the British Rolls Royce Nene engines for the licensed American version of the Nene, the Pratt & Whitney J42-P-6 engine.

In January 1947, a mock-up of the proposed fighter was inspected and, with some minor modifications, accepted by the Navy. In the following month, metal was cut on the prototypes. The US Navy began negotiations with Grumman to acquire an early batch of 30 F9F-2s (BuNos 122560-122589) despite prototype flight testing being months away. Such was the interest in the new fighter, which Grumman had now named Panther, to continue with the wild feline theme of its earlier aircraft.

The big day was getting near. The first XF9F-2 powered by a 5,000 lb. thrust Nene was rolled out at Bethpage in mid-November 1947. High-speed taxiing trials commenced on November 20.

Finally, on November 21, the XF9F-2 took off on its maiden flight with Corwin Henry 'Corky' Meyer at the controls. The test was non-eventful, with Meyer landing at Idlewild International Airport as a safety measure due to the airport's longer runway than Bethpage, which had a runway only 5,000 feet long. The Idlewild landing resolved some questions concerning the braking capability of the aircraft, and Meyer promptly took off again and landed back at Bethpage.

Two US Navy Grumman F9F-2 Panthers from the famous Fighter Squadron 51 (VF-51) 'Screaming Eagles' during a sortie over Korea, ca 1951. VF-51 was assigned to Carrier Air Group 5 (CVG-5) aboard the aircraft carrier USS Essex (CV-9) for a deployment to Korea from June 26, 1951 to March 25, 1952. The lead plane's pilot (S-107) is Lt. JG George Russell, while in plane S-116 the pilot at the controls is none other than the first man on the moon, Neil Armstrong. (US Navy, Naval History and Heritage Command, John Moore)

Three weeks after the inaugural flight of the XF9F-2, the first order for 47 production F9F-2 carrier-based fighter-bombers with the Pratt & Whitney J42-P-6 engine, the licensed version of the Nene, was awarded to Grumman. It was followed immediately by an order for 54 F9F-3 fighter-bombers, with the Allison J33-A-8 engine, 4.600 lb. s.t. (2,086kg). The Allison engine was considered a backup alternative to the Pratt & Whitney J42-P-6 in the event that Pratt & Whitney might experience some delays and not be able to deliver the engines or not be able to ensure sufficient production of the J42 engines. The second XF9F-2 was completed and flown in the spring of 1948 with no issues. After the start of mass production of the F9F Panther, the third prototype, the XF9F-3 with the Allison J33 engine, took its first flight on August 16, 1948.

In November 1948, barely a year after the XF9F-2s first flight, the first models of the F9F-2 versions were completed simultaneously. The Pratt & Whitney J42 engine's production was proceeding without significant problems. Moreover, the successful performance of the Pratt & Whitney engine meant all the Allison F9F-3s were to be re-engined from February 1950 with the Pratt & Whitney J42-P-6, thus becoming F9F-2s. Subsequently, the Pratt & Whitney J42 would power all F9F-2s in Navy service, a total of 567 aircraft.

USMC F9F-2 (BuNo 123451) taking a break from the war in Korea. Delivered to the United States Navy in June 1950 and it went through the following squadrons 1950: VMF-312. 1950-51: VMF-311. 1952: VMF-311 as WL-2. 1952: VF-821. VF-63. VF-653. 1953: VF-151. 1955: ATU-223. ATU-202. In January1957, it was finally put into storage at NAF Litchfield Park, AZ. (National Archives)

Pratt & Whitney had quickly prepared the Nene for American manufacturing. The company was churning out its J42-P-6 engines as fast as Grumman was building airframes. The P&W J42-P-6 was the Panther's first engine, but the newer 5,000 lb. thrust, 5,750 pounds with afterburner, Pratt & Whitney J42-P-8 quickly replaced it.

The Navy's original deal with Grumman was for 437 F9F-2s. The first delivery to an operational unit was to 'VF-51' 'Screaming Eagles' at NAS North Island, San Diego, on August 5, 1949. That squadron pioneered the use of fighters in naval service with the North American Aviation FJ-1 Fury. Most of the F9F-2s went to Navy squadrons, but some Marine Corps Squadrons were also equipped with F9F-2 Panthers.

An interesting story on the Panther surfaced in June 1950. Inside the Pentagon, the staff was buzzing over a report from official 'judges' of 'Operation Portrex' regarding the massive joint maneuvers held on Vieques Island in Puerto Rico. According to the judges, the official scores were well in favor of the Navy. At the end of the training phase, the Marine squadron, 'VMF-115' 'Silver Eagles' or 'Able eagles' flying F9F-2s hypothetically destroyed ninety Air Force F-84 Thunderjet fighter bombers with a loss of only nine F9F Panthers. The Air Force was stunned. The Navy and its F9Fs were showing the way to fight air battles. The real proving ground, however, was going to be Korea.

US Navy Grumman F9F-3 Panther (BuNo 122562) was operated by the Naval Air Test Center at Patuxent River, Maryland (USA). This aircraft was fitted with an experimental electro-hydraulically driven Emerson Aero X17A roll-traverse turret housing four .50 caliber machine guns, which replaced the standard 20 mm cannon. The guns could be directed at any angle from directly forward to 20 degrees aft, and the gun mount could roll 360 degrees. Note the four T Emerson Aero X17A guns pointing downward. The volume required for the fire control system avionics, and the sheer weight of the turret, made it impractical for single-seat fighters and the program was cancelled in early 1954. (USN photo National Naval Aviation Museum)

US Navy Grumman F9F-2 Panther of Fighter Squadron 112 (VF-112) 'Fighting One Twelve' in flight. VF-112 was assigned to Carrier Air Group 11 (CVG-11) for four deployments to Korea aboard the aircraft carriers USS Valley Forge (CV-45) and USS Philippine Sea (CV-47) 1950 - 1952. (National Archives)

F9F-2P

The Navy was interested in procuring a jet-powered photo recon aircraft that could stay in touch with the Navy's jet fighters. They found the answer in their own 'backyard.' The solution was to use one of their F9F-2s by removing its guns and replacing them with weapons of peace, cameras. Additionally, windows were fitted underneath the nose and on each side of the nose area of the Panther. Inside the cockpit, a viewfinder replaced the gunsight, and the camera controls were placed on the left console. Access to the camera bay was through a hatch on top of the nose. Although the Panther's nose section was not lengthened, the aircraft kept its measurements intact even after all these modifications. The Navy now had its very own jet recon aircraft.

Grumman F9F-2P Panther photo recon aircraft is parked on the hangar deck of USS Bon Homme Richard (CVA-31), during operations off Korea, on November 15, 1952. The plane, which appears to be BuNo 123700, bears markings for 85 missions and is nicknamed 'LIFE', after the popular magazine. Two of the windows for the cameras can be seen under the nose, and left side of the nose of the aircraft.
(Naval History and Heritage Command)

F9F-2 & F9F-3 SPECIFICATIONS

TYPE	POWERPLANT	DIMENSIONS WINGSPAN / LENGTH / HEIGHT	WEIGHTS Empty / Combat / Gross / Max. t.o.w.	SPEEDS Cruise / Max Sea Level / At 22,000 ft / At 35,000 ft / Landing Speed	SERVICE CEILING / RANGE
F9F-2	Pratt & Whitney J42-P-6, 5,750 lb. s.t. (2,601 kg)	35ft 3in (10.74m)/ 37ft 3in (11.35m)/ 11 ft 6in (3.50m)	9,303lb. (4,220kg)/ 14,235lb. (6,457kg)/ 16,450lb. (7,462kg)/ 19,494lb. (8,842kg)	487mph (784kph)/ 575mph (925kph)/ 545mph (877kph)/ 529mph (851 kph)/ 105mph (169kph)	44,600ft (13,594m) / 1,353miles (2,177km)
F9F-2 WEAPONS	Four forward-firing Mk-3 20mm cannon, 2,000lb (907kg) bombs, or six 5-in. rockets				

F9F-3 Powerplant:
(ORIGINALLY) Allison J33-A-8 engines 4,600 lb. s.t. (2,086kg). (LATER) Pratt & Whitney J42-P-6, 5,750 lb. s.t. (2,601 kg)

All Dimensions, Weights, Speeds, Service Ceiling, Range, and weapons are in accordance with the F9F-2 variant.
(Post P&W J42-P-6 installation)

F9F-4

The F9F-2 and F9F-3 had gone into production without significant alterations. Out of those two variants, the F9F-2 was manufactured in more substantial quantities. Engineers at Grumman then shifted their focus to improving the fundamental design of the Panther while the Navy was thinking of methods to increase the power of its jets. The new variant, the F9F-4, gave the Navy hope that this would be the winner. The new F9F-4 version differed physically from the older Panthers in a number of ways. For example, it had a more extended nose cone and a higher, more pointed tail. Furthermore, Allison fitted the Navy's F9F-4s with its more powerful Allison J33-A-16 engine with 6,950 pounds of thrust (with water injection). The J33-A-16 enhanced the F9F-4s speed quite dramatically. Seventy-three F9F-4s were consequently equipped with the Allison J33. However, the Navy had jumped the gun. The Allison J33-A-16 was found to be a problematic engine that was the bane of the F9F-4. Sure, the J33 increased the F9F-4's speed and performance, but the engine's inner workings were not in good shape. There were severe issues with the bearings on the J33-A-16 engines. Eventually, the F9F-4 prototype, which was actually an F9F-2 conversion (s/n 123084), made its first flight with Allison's J33-A-16 engine on July 6, 1950, months behind schedule. However, due to the engine's ongoing issues, and mediocre operational characteristics, most of the 109 F9F-4s were upgraded to the F9F-5 variant with the Pratt & Whitney J48-P-6A or P-8 engines.

USMC F9F-4 Panther, (BuNo 125183), of Fighter Attack Squadron 311 (VMF-311) 'Tomcats'. VMF-311 was redesignated VMA-311 in 1957. On October 15, 2020, VMA-311 was deactivated, however, the squadron was planned to be reactivated as Marine Fighter Attack Squadron 311 (VMFA-311) 'Tomcats', in spring 2022. The F9F-4 above is currently preserved at Pima Air and Space Museum, Tucson, AZ. (Andrew Thomas)

F9F-4 Specifications

TYPE	POWERPLANT	DIMENSIONS WINGSPAN / LENGTH / HEIGHT	WEIGHTS Empty / Combat / Gross / Max. t.o.w.	SPEEDS Cruise / Max Sea Level / At 35,000 ft / Stalling Speed	SERVICE CEILING / RANGE
F9F-4	Allison J33-A-16, 5,850lb s.t. (2,653kg)	35ft 3in (10.74m)/ 38ft 10.5 in (11.83m)/ 12ft 4in (3.75m)	10,042 lb. (4,555kg)/ 15,264 lb. (6,924kg)/ 17,671 lb. (8,016kg)/ 21,250 lb. (9,639kg)	495mph (797kph)/ 593mph (954kph)/ 547mph (880kph)/ 131mph (211 kph)	44,600ft (13,594m)/ 1,175miles (1,891km)
F9F-4 WEAPONS		4x20mm Cannon, plus underwing weapon load of up to 2000 lb.			

F9F-5

The first F9F-5, also a converted F9F-2 airframe, first flew on December 21, 1949, beating the earlier variant, XF9F-4, which made its first flight seven months later, in July 1950. Deliveries of the F9F-5 began on November 5, 1950. The F9F-5 was the final of four Panther variants built and was made in higher numbers than any previous variant. A few months after the F9F-5s inaugural flight, the Korean War broke out, prompting intense manufacturing of the '-5' between 1950 and 1952. The F9F-5s were equipped with uprated Pratt & Whitney J48-P-6s or P-8s, which generated 7,000 pounds of thrust with water/alcohol injection, and 6,250 pounds maximum dry thrust. The '-5' version was the definitive Panther model. With the Korean War heating up, most US Navy squadrons equipped with F9F-2s converted to the later F9F-5 model.

As had been the case with the F-2 and F-3, the F-4 and F-5 were identical except for the engines. Both versions were characterized by a slightly lengthened rear part of the fuselage and had a higher vertical tail and tail rudder. Likewise, the folding angle of the wings increased from 62° to 59°, which resulted in somewhat more bulk during storage in the hangars of the aircraft carriers. A photographic reconnaissance version, the F9F-5P, was also developed, with a modified nose housing a battery of cameras for planimetric photos (photos taken showing positions on the ground, such as roads, buildings, factories, water, vegetation, bridges, railroads, etc.). Featuring the basic F9F-5 airframe, the F9F-5P also had a longer nose. A total of 36 photographic F9F-5Ps were built. As for the F9F-5, 655 were built. A total of 1,388 Panthers left the assembly lines when the final F9F-5 was delivered to the US Navy in December 1952.

F9F-5 Specifications

TYPE	POWERPLANT	DIMENSIONS WINGSPAN / LENGTH / HEIGHT	WEIGHTS Empty / Combat / Gross / Max. t.o.w.	SPEEDS Cruise / Max Sea Level / At 35,000 ft / Stalling Speed	SERVICE CEILING / RANGE
F9F-5	Pratt & Whitney J48-P-4/6A with water injection, 6,250lb. s.t. (2,835kg), 7,000lb. a.b. (3,175kg) / J48-P-8, 7,250lb s.t. (3,288kg)	35ft 3in (10.74m)/ 38ft 10.5 in (11.83m)/ 12ft 4in (3.75m)	10,147lb. (4,603kg)/ 15,359lb. (6,967kg)/ 17,766lb (8,059kg)/ 18,721 lb. (8,492kg)	481mph (774kph)/ 604mph (972kph)/ 543mph (874kph)/ 131mph (211kph)	42,800ft (13,045m)/ 1,300mi (2,092km)
F9F-5 WEAPONS		4x20mm M-3 Cannon, each with 190 rounds. Up to six 500-pound bombs or two 1,000-pound bombs could be carried. Other combinations were six 5-inch High-Velocity Aircraft Rockets (HVAR), eight 100-pound bombs, or eight 250-pound bombs.			

F9F-5 Panthers of fighter squadron VF-114 'Executioners' (later 'Aardvarks') lined up on deck of the aircraft carrier USS Kearsarge (CVA-33) in 1955. VF-114 was assigned to Carrier Air Group Eleven (CVG-11) during deployment from October 7, 1954 to May 12, 1955 to the Western Pacific. Squadron VF-114 was disestablished in 1993. (US Navy Photo, Collection of National Naval Aviation Museum)

This is F9F-5 Panther (BuNo 126066) in flight near a snowcapped volcano. This is the personal aircraft of the Commander of Carrier Air Group 2 (CVG-2). The photo was taken about 1953-54 when fighter squadrons VF-63 'Fighting Redcocks' and VF-64 'Free Lancers' flying F9F-5s were assigned to CVG-2 for a deployment to the Western Pacific aboard the aircraft carrier USS Yorktown (CVA-10). VF-63 was redesignated VA-63 in March 1956. Then, on July 1,1959, it was redesignated again as VA-22. Finally, VA-22 was redesignated VFA-22 on May 4, 1990. VF-64 was redesignated VF-21 on July 1,1959. (US Navy Photos, Collection of National Naval Aviation Museum)

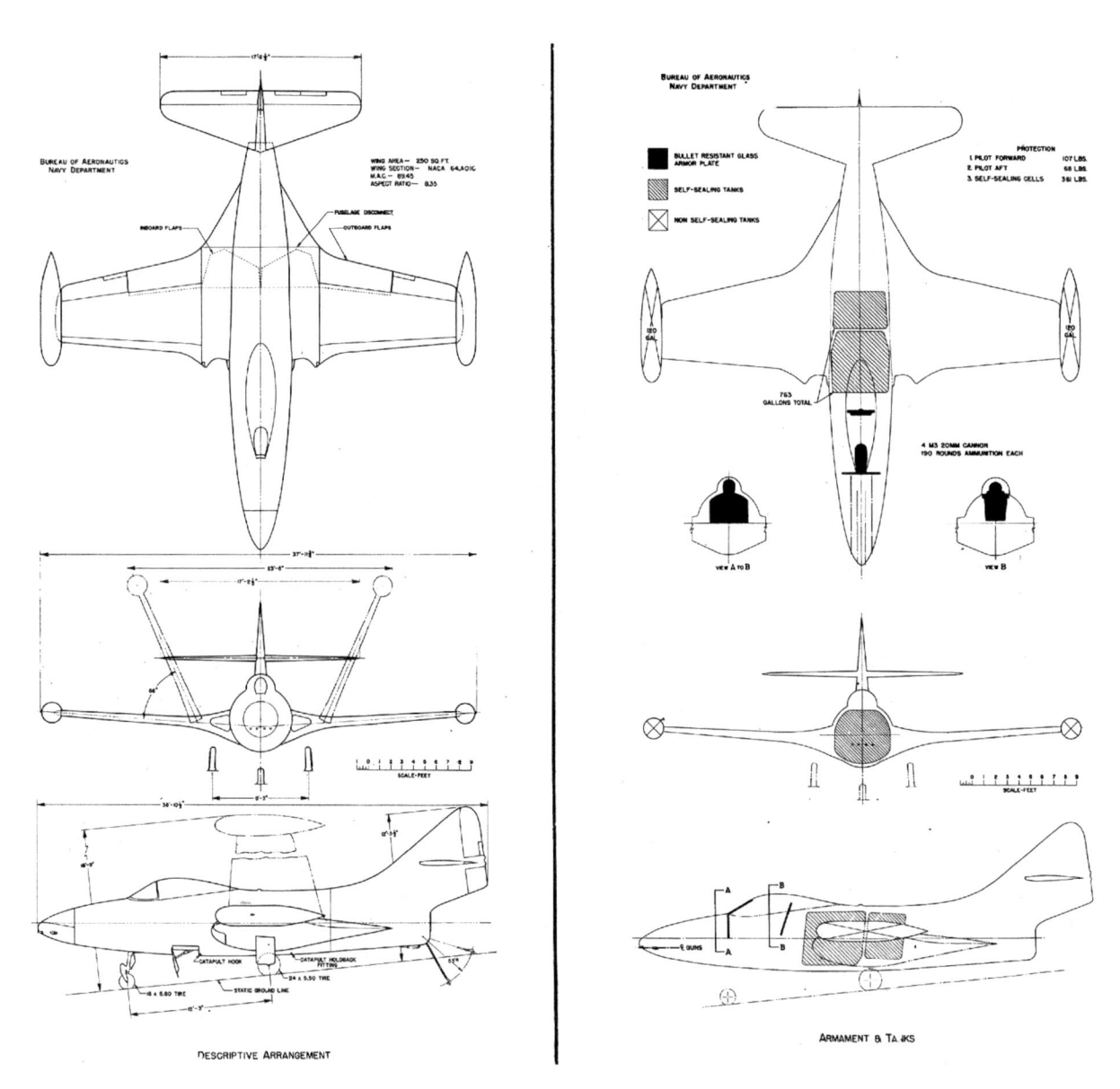

GRUMMAN F9F-5-5P PANTHER - STANDARD AIRCRAFT CHARACTERISTICS (Graphics: US Navy)

Grumman F9F-2 Panther, BuNo 122560, at the Langley Aeronautical Laboratory at Hampton, Virginia. It was originally built as a F9F-3, but was refitted with a Pratt & Whitney J42 turbojet power plant, and became an F9F-2. It was then sent to NACA in 1951. During its time with NACA/NASA, it was used to test various autonomous flight control systems. The aircraft served long enough at Langley NACA/NASA to witness the change from the NACA to NASA, on 1 October 1958. (NASA)

THE LANGLEY-BASED F9F-2 PANTHER

After a stint at NAS Norfolk, Virginia, Grumman F9F-2 Panther, BuNo 122560, built initially as an F9F-3, was re-engined with a Pratt & Whitney 42 turbojet engine and became a '-2'. After that, it was assigned to the National Advisory Committee for Aeronautics (NACA), the predecessor to NASA, situated at Langley, Virginia, on January 5, 1951. Once The Panther joined NACA, it was registered as NACA 215.

NACA Langley Aeronautical Laboratory launched an initiative in 1952 to investigate different features of the fly-by-wire system, including using a side-stick controller. As a result, flight testing began in 1954 with what was possibly the first jet-powered fly-by-wire research aircraft. The chosen aircraft was the modified former US Navy Grumman F9F-2 Panther carrier-based jet fighter, now operating as a NACA research aircraft.

The NACA effort aimed to assess several automatic flight control systems based on rate and normal acceleration feedback. Its secondary goals included evaluating the usage of the fly-by-wire system with a side stick controller for pilot inputs. The original F9F-2 hydraulic flight control system and its mechanical linkages were preserved, with NACA creating an auxiliary flight control system based on a fly-by-wire analog principle. At the end of the right ejection seat armrest was a tiny 4-inch high-side stick controller. The controller was pivoted at the bottom and was used for both lateral (roll) and longitudinal (pitch) control. Full stick deflection needed only 4 pounds of pressure. In addition, the electrically driven system eliminated control friction, typically present in a hydromechanical system (the mechanics of fluids, especially water).

Furthermore, the aircraft's fuel system was changed so the fuel could be pushed rearward to destabilize the airplane by shifting its center of gravity backward. In addition, a steel container was mounted at the lower rear of the aircraft's fuselage, with 250 pounds of lead shot loaded upon it, severely destabilizing the plane. In an emergency, the lead shot could be rapidly ejected to stabilize the aircraft once again. After all the testing was done, the modified F9F-2 was flown by fourteen pilots.

Using only the side-stick controller, the pilots conducted takeoffs, stall approaches, aerobatics, and rapid precision maneuvers, including air-to-air target tracking, ground strafing runs, precision approaches, and landings. The test pilots rapidly became accustomed to flying with the sidestick and considered it both comfortable and natural. The F9F-2 went through many ordeals during the testing by NACA of the fly-by-wire system. It has to be said that the tests conducted with the Panther greatly assisted the NACA and its successor NASA in gaining vital knowledge, which would improve the avionics in future aircraft designs. The X-15 was one of those aircraft. The NACA's test data derived from the F9F-2 fly-by-wire experiment was used to develop the side-stick controllers in the North American X-15 rocket research aircraft with its adaptive flight control system. We could confidently say that the Grumman F9F-2 Panther, serial number 122560, played a more significant part in the enhancement of aviation technology than any other Panther or Cougar that came before or after it.

After serving the NACA and its successor NASA, the Grumman F9F-2 was struck off charge at Langley AFB in Virginia on January 14, 1958. It was scrapped on July 7, 1960.

THE F-9F IN THE 'FORGOTTEN WAR' KOREA, 1950 – 1953

Korea is a nation that has had more than its fair share of occupations. Larger neighbors have invaded, influenced, and battled over Korea for a significant portion of its history. Throughout its 2,000-year existence, it has experienced around 900 invasions. From 1231 through the beginning of the 14th century, Korea was occupied by the Mongols. After the Mongols left the scene, it was the turn of the Chinese government and Chinese rebel armies to ravage Korea regularly. Then came the Japanese. China's defeat by Japan during the Sino-Japanese War in 1894 secured Japan's suzerain rights over Korea (i.e., Japan controlled Korea's foreign affairs, while Korea only managed its domestic affairs). Sixteen years later, in 1910, Japan formally annexed the country. The Japanese colonial era was characterized by tight control from Tokyo and its ruthless efforts to erase the Korean language and culture. Such is the fate of a nation placed under the giant shadow of China on one side and Japan on the other.

After the Second World War. Following the liberation of Korea from Japan, the United States occupied Korea, a size roughly equivalent to modern-day South Korea, for three years. A lot of uncertainty and confusion marked this period. The lack of a clearly defined US strategy toward Korea, the escalation of US-USSR hostilities, and the polarization of Korean politics between the left and right were significant contributors to this challenging position. Korea was split into two at the 38th Parallel.

The Communist side was in the North, and the Republican side was in the South. The uncertainty prevailed until 1947, for a brief time, when the United Nations General Assembly recognized Korea's claim to independence. On August 15, 1948, the Republic of Korea (South Korea) was proclaimed, and preparations for the establishment of a government, and the withdrawal of occupation forces, both American and Soviet, were made. In June 1949, the Americans returned to the United States. With the US out of the picture, confidence was high. The Communists were getting ready to move in.

Less than a month after the Republic of Korea was proclaimed, the Democratic People's Republic of Korea (North Korea) was established under Premier Kim Il Sung. He claimed control over the whole country based on elections held in the North and 'allegedly' in the South of the country.

In February 1948, the Korean People's Army (KPA) was formed in North Korea by Korean communist guerrillas who had previously served with the Chinese People's Liberation Army but were 'guided' by Soviet soldiers.

North Korean forces totaled between 150,000 and 200,000 soldiers by June 1950, divided into ten infantry divisions, one tank division, and one air force division. In early 1950, Soviet equipment, including various types of automatic weapons, T-34 tanks, and Yak fighter aircraft, flooded North Korea. The highly motivated, well-trained North Korean troops were going to confront a South Korean army that lacked tanks, heavy artillery, and combat planes. It had a coast guard of 4,000 soldiers and a police force of 45,000 men. Basically, they were as unprepared for War as the Poles were when the Germans unleashed Blitzkrieg on them to start World War II. For the South Koreans, the invaders this time were not the Mongols, Chinese, or Japanese; it was their fellow citizens taking aim. The die was cast, and the scene was set for the first significant War after the end of World War II. D-Day for the North Koreans was set for June 25, 1950. Even though the South Koreans knew that the Communists would invade the South sometime, they were ill-prepared for what would transpire.

The events that followed the June 25 invasion demonstrated North Korea's military strength and the validity of its entire invasion strategy. The army of South Korea was overpowered, and Seoul fell in three days. By early August, South Korean soldiers were constrained to a zone 87 miles long and 56 miles wide in the peninsula's southeastern corner, named the Pusan Perimeter. The North Korean army controlled the remainder of the area.

On the day of the invasion, the US requested an emergency meeting of the UN Security Council, which then passed a resolution condemning North Korean aggression, ordering the withdrawal of North Korean troops above the 38th Parallel, and urging all members to assist the UN in carrying out the resolution. However, the Soviets had boycotted the meeting in protest of the United States' continued backing of Chinese nationalists, and their absence prevented them from vetoing the proposal.

On June 27, President Harry S. Truman declared that he had ordered the Far East's sea and air forces to provide assistance and cover for the South Koreans, as well as the Seventh Fleet, to take measures to avoid an invasion of Formosa (Taiwan). Later that night, the UN Security Council voted a resolution urging all of its members to support South Koreans in repelling the North Korean assault.

THE GRUMMAN F9F PANTHER JOINS THE WAR IN KOREA

Looking at the first battle in the Korean War by US Naval forces, it was by luck that the US Navy had an aircraft carrier anywhere near Korea in 1950. In 1946, US postwar strategy necessitated the deployment of two aircraft carrier groups in the Western Pacific, but

US Navy Grumman F9F-2 Panther (BuNo 127210) of Fighter Squadron 24 (VF-24) 'Corsairs' flies over ships of Task Force 77 with its arresting hook extended before landing aboard the aircraft carrier USS Boxer (CV-21) on July 4, 1952 during operations off the coast of North Korea. Note the two aircraft carriers in the sea, on the right of the photo. From February 8 to September 26, 1952, VF-24 was assigned to Carrier Air Group 2 (CVG-2) aboard the Boxer for deployment to Korea. Later, on July 26, 1953, the F9F-2, 127210, ditched off the USS Boxer while serving with VF-151 'Black Knights.' (National Archives)

budgetary constraints prevented this from occurring. More astonishing was the fact that there was not a single US aircraft carrier in the whole of the Western Pacific at the beginning of 1950, only six months before the outbreak of the Korean War. Then, from 1950, US policy mandated the deployment of single carrier units on six-month service deployments in the Western Pacific.

Subsequently, Admiral Struble's Seventh Fleet Striking Force, 'Task Force 77', was deployed to the Western Pacific on May 1, 1950, for a six-month stint, ending on December 1, 1950. Task Force 77 consisted of a 'carrier group' with one carrier on duty; 'USS Valley Forge' (CV-45), a 'support group' with one cruiser, and a 'screening group' with eight destroyers

Eight days after the beginning of hostilities, on July 3, 1950, the carrier USS Valley Forge went into action under the UN flag. USS Valley Forge's 86-plane complement included two jet fighter squadrons, 'VF-51' and 'VF-52', with 30 Grumman F9F-3 Panthers and 40 pilots. It also had two piston-engine fighter squadrons with World War II era Vought F4U-4B Corsairs 'VF-53' and 'VF-54', and a piston-engine assault squadron with 14 Douglas AD-4/Q Skyraiders 'VA-55'. Aside from these five squadrons, the group also had another squadron of F4U-5N Corsairs and AD-3N Skyraiders, 'VC3DetC', configured explicitly for night photographic 'expeditions' and radar missions. USS Valley Forge's F9F Panther squadrons and the other squadrons were ready for battle.

At 5 am, on July 3, Valley Forge's Carrier Air Group 5 (CVG 5) unleashed its first strikes in the Korean conflict. First, the carrier launched combat and antisubmarine patrols. Then, at 5.45 a.m., the British carrier in the Group 'HMS Triumph' launched 12 Fairey Fireflies and nine Supermarine Seafires for an attack on the airfield at Haeju. The Firefly was a World War II-era carrier-borne fighter and antisubmarine aircraft. The Seafire was the naval version of the famed World War II fighter, the Spitfire. Then at 6 am, Valley Forge commenced the launching of its strike group. First, 16 F4U-4B Corsairs, each armed with eight 5-inch rockets, and twelve AD4/Q Skyraiders, each carrying one 1,600-pound bomb, were launched for an attack on Pyongyang's airfield. Once the propeller-driven attack aircraft got a good head start, Valley Forge launched eight F9F-3 Panthers, whose greater speed would bring them over the target area first. After that, the attacks would be concentrated on North Korean lines of communication, with railroad bridges, rail yards, airfields, and roads on the list of targets.

As planned, the F9F-3s arrived first in the target area and flew over the North Korean capital without encountering significant resistance. Nine Korean People's Army Air and Anti-Air Force (KPAAF) aircraft were reportedly destroyed on the ground, while two Yak-9s were shot down in air battles with the F9F-3 Panthers. One other Yak-9 was damaged. The unexpected presence of USAF jet aircraft over 400 miles away from the nearest American air base alarmed the North Koreans. Without a doubt, the Panthers' appearance so far north prevented a significant deployment of KPAAF aircraft to North Korean air bases in case they get destroyed by the American jets.

Soon after the F9Fs finished their destruction of the meek KPAAF resistance, the F4U-4B Corsairs and AD4/Q Skyraiders flew in and fired their rockets at the airfield's hangars and fuel tanks. At Pyongyang and Haeju, enemy antiaircraft resistance was light, and no US Navy aircraft sustained significant damage. In the afternoon, planes from HMS Triumph flew in for a second raid, while USS Valley Forge launched a second attack against the Pyongyang marshaling yards and the Taedong River bridges. There were reports of significant damage to locomotives and train stock, but the bridges withstood the onslaught.

North Korea's Air Force, the 'Korean People's Army Air, and Anti-Air Force' (KPAAF), did a good job of wreaking havoc on the South Korean forces in the early days of the War. However, in any meaningful sense, the KPAAF's strengths were limited. After the commencement of hostilities on June 25, estimates of the number of aircraft in the KPAAF ranged between 75 and 130 planes, all of which were vintage WWII propeller types, mainly Ilyushin Il-2 and Ilyushin Il-10 ground attack aircraft, and Yakovlev Yak-9 fighters. They were no match for the US Navy's fighters and attack aircraft.

The first day of battle was a big success for Admiral Struble's Seventh Fleet Striking Force, 'Task Force 77', and USS Valley Forge. All the squadrons completed their tasks as expected. North Korean airfields were severely damaged, railway lines and rail yards felt the effects of the massive bombing attacks by the Navy aircraft, and many roads were rendered unusable. However, that was not the biggest news of the day. Instead, it was the two Navy pilots of Squadron VF-51, Lt. JG Leonard H. Plog and Ens. Elton W. Brown Jr. Each shot down one KPAAF YAK-9 aircraft and destroyed two more Yak-9s on the ground. LTJG Plog was credited with the first US Navy kill of the War, and his action was the first time in history a US Navy carrier-based jet fighter had shot down an enemy aircraft.

There was a temporary lull in fighting until Tuesday, July 18. On that day, Valley Forge (CV 45) and British carrier Triumph (R 16) resumed operations with attacks on North Korean airfields, railways, and industrial targets at Hamhung, Hungnam, Numpyong, and Wonsan.

USS Valley Forge in July 1950. Flight deck tractors tow Grumman F9F Panther fighters forward on the carrier's flight deck, in preparation for catapulting them off to attack North Korean targets. (Naval History and Heritage Command).

Under attack by aircraft from Valley Forge (CV-45) on 18 July 1950. Smoke from this attack, which reportedly destroyed some 12,000 tons of refined petroleum products and much of the plant, could be seen sixty miles out at sea. (Naval History and Heritage Command).

Their aircraft, including the F9Fs, severely damaged the Wonsan oil refinery. The carrier aircraft flew interdiction strikes deep behind enemy lines and close support missions as needed for the remainder of the month. At the same time, USS Valley Forge and Triumph shifted completely around the peninsula from the Sea of Japan to the Yellow Sea to relieve the pressure on UN troops retreating toward the southern region of Pusan, in South Korea.

The USS Valley Forge was the only aircraft carrier fighting in Korea in May, June, and July, but assistance was on its way. With the commencement of hostilities in Korea, all plans and schedules were thrown out the window. USS Philippine Sea (CV-47) with Air Wing 11 (CVG-11) set sail from San Diego on July 6 for Pearl Harbor, where she arrived on July 14 to begin a ten-day phase of intensive training exercises. After that, she set sail again, bound for Buckner Bay, Okinawa, where she arrived on August 1. At Okinawa, USS Philippine Sea, an Essex class carrier, reported to Commander Seventh Fleet at Buckner Bay. However, before the arrival of the USS Philippine Sea, Badoeng Strait (CVE-116) landed in Yokosuka, Japan, on July 22, carrying members of the 1st Marine Aircraft Wing. Next, on July 26, Sicily (CVE-118) also arrived in Yokosuka with a load of ammunition. Then on July 31, USS Valley Forge, also an Essex class carrier, anchored at Bruckner Bay. The following day Valley Forge's crew was relieved to see the arrival of the USS Philippine Sea. Now with two Essex-class aircraft carriers in South Korea, things would get easier for USS Valley Forge.

The Captain of the USS Philippine Sea was Willard Kinsman Goodney. This deployment was the carrier's first of three in the Korean conflict for the USS Philippine Sea that would endure a grueling nine months (From August 7, 1950, to April 7, 1951). The ship carried Carrier Air Group Eleven (CVG 11). The aircraft carrier's first Commander Vogel was killed in action on August 19, 1950, barely two weeks after the arrival of the USS Philippine Sea to the war zone. Commander Weymouth replaced him. Regarding the squadrons on the carrier, they were as follows: VF-111 and VF-112: F9F-2 Panthers, VF-113 and VF-114: F4U-4B Corsairs, VA-115: AD-4/Q Skyraiders. In addition, the USS Philippine Sea also carried typical detachments; VC-3 Det. 3: F4U-5N radar-equipped Corsairs and AD-4N night attack Skyraiders, VC-11 Det.: AD-4W airborne early warning Skyraiders, VC-61 Det. 3: F4U-4P photo recon Corsairs, and HU-1 Det. 3: HO3S-1 helicopters.

July was a bad month for the Navy's aircraft. Two F9Fs, three F4Us, and a helicopter crashed into the ocean, and on July 22, an AD Skyraider crashed and caught fire, killing its pilot. Most downed personnel, however, were fished out of the water by screening ships. One pilot was recovered 80 miles from the allied shipping zone by HMS Triumph's amphibious aircraft. Meanwhile, an Army helicopter picked up another pilot whose plane had gone down due to fuel exhaustion behind enemy lines.

On August 21, the two Essex carriers upped the ante. Aircraft flying from Valley Forge (CV-45) and the Philippine Sea (CV-47) demonstrated the intensification of the air war by performing 202 sorties in a day over the Pyongyang region of North Korea. Again, the F9Fs played a pivotal role in the attacks.

On September 6, 1950, another Essex class carrier, USS Leyte (CV-32), sailed from the Atlantic Fleet and arrived in the Korean War theater on October 9. Captain Thomas Upton Sisson was on the bridge of the Leyte. The Carrier Air Group was CVG-3. Its squadrons consisted of the following aircraft: VF-31: F9F-2 Panthers, VF 32: F4U-4 Corsairs, VF-33: F4U-4 Corsairs, VA-35: AD-3 Skyraiders. Detachments were as follows: VC-4 Det 3: F4U-5N Corsairs, VC-33 Det 3: AD-4N Skyraiders, VC-12 Det 3: AD-3W early-warning radar picket Skyraiders, VC-62 Det 3: F4U-5P long-range photo-recon Corsairs, and HU-2 Det 3: HO3S-1 helicopters.

THE INCHON LANDING

The Battle for Inchon began on September 15, 1950, and ended on September 19. The landing at Inchon was undertaken by the US and South Korean soldiers at the port of Inchon, near the capital of South Korea. The operation was a bold one that involved some 75,000 troops and 261 naval vessels and led to the recapture of the South Korean capital of Seoul. The famed US General Douglas MacArthur orchestrated the Inchon landing, which was accomplished under extremely challenging circumstances. The battle ended a string of victories by the North Korean People's Army (KPA) and forced the KPA to escape in disorder up the Korean peninsula. Moreover, it represented a significant victory and strategic reversal for United Nations Command (UN).

Following the successful invasion at Inchon in late September, Admiral Struble's Seventh Fleet Striking Force, Task Force 77, turned its attention to interdiction operations beyond the Main Line of Resistance (MLR) into North Korea, seeking various targets. Valley Forge's Carrier Air Group 5 (CVG 5) spotted a North Korean truck convoy in the open six miles east of Inchon, at Taejong. The aircraft of CVG 5 went straight for them and destroyed 87 vehicles. The US Navy and The USAF had freedom of the skies. The winds of fortune were blowing in favor of the US and its UN allies.

THE LUCKY PILOT

During the Task Force 77 interdiction operation, an F9F-2B Panther pilot from the USS Philippine Sea's 'VF 111' squadron, known as the 'Sundowners,' conducted the first combat ejection from a jet fighter in battle. While on a ground attack mission over North Korean territory, LT Carl C. Dace was hit by Antiaircraft (AA) fire while strafing at 460 mph at an altitude of 2,000 feet. The AA hits knocked out the Panther's fuel system and swiftly reduced his fuel supply to just 50 pounds. With throttle full on, he could only get 85 % power. He decided to climb up to 6,000 feet, where he pulled the pre-ejection lever to dislodge his canopy. It moved aft but refused to jettison due to battle damage. In the end, he pushed it off with his hand.

Dace then drew the face curtain down to protect his face from debris before being catapulted from the jet, which was traveling at roughly 230 mph. According to his wingman, the small stabilizing drogue parachute on the cockpit seat did not deploy, causing the seat and Dace to tumble violently through the air. Dace then let go of the face curtain with one hand to unfasten his safety belt, which held him in his seat. This action resulted in the free end of the curtain flapping in the slipstream while the nylon rope handle hit him in the eye. It would have pulled back if he had let go of the curtain with both hands. Unfastening his belt, he kicked the seat free and pulled his ripcord. One panel was torn out of the chute, but it functioned despite the damage, and he descended into the sea.

He landed in the water, breaking out his PK 2 Pararaft kit containing a one-man life raft and other survival accessories. He got on the raft and waited for help. After seven long hours in the water, support finally came in the form of a destroyer, which picked him up and took him back to safety.

MiG-15

In late October 1950, Communist Chinese soldiers were spotted in clustered groupings along the front line. China asserted that these troops were only 'volunteers' sent to North Korea to help their socialist brothers. However, at the beginning of November, it was evident that Chinese troop infiltration along the front was increasing and posed a grave threat. To deter Chinese infiltration, the Navy intended to destroy the Yalu River bridges between China and North Korea. Task Force 77 was entrusted with handling the situation on November 8. The carrier air groups were ordered to attack the Yalu bridges from Sinuiju across the peninsula to Hyesanjin, some 200 miles upstream.

Lt. Cmdr. William T. Amen of VF-111 'Sundowners' from the USS Philippine Sea (CV 47), scored the Navy's first MiG-15 kill in Korea on November 9, 1950.
(Naval History and Heritage Command)

The attack force consisted of F4U Corsairs, AD Skyraiders, and F9F Panthers as top cover. Chinese Communist Air Force MiG-15 jet fighters based at Antung Airfield in Manchuria intercepted the US Navy aircraft, which aimed to destroy the bridge at Sinuiju. Amid this confrontation, Lt. Cmdr. William T. Amen, of VF-111 'Sundowners', shot down one of the attacking MiG-15s, the first jet kill recorded by a US Navy pilot.

The Mig-15 kill by Lt. Cmdr. Amen boosted morale for the allied forces. However, the euphoria did not last long. Reports of a Chinese buildup in Manchuria, the crossing of Chinese 'volunteers' into North Korea, and combative declarations from Beijing had alarmed Allied strategists. With the inclusion of Chinese forces in the Korean conflict, things could quickly get out of hand.

On top of that was the introduction of the new Soviet MiG-15 jet in the War. As their flight skills improved, the Chinese pilots showed more aggressiveness. Moreover, it became pretty clear that the new Soviet-made MiG fighter vastly outperformed any aircraft deployed by the United Nations forces in Korea. It could fly faster, dive faster, climb more quickly, and even turn round more swiftly than the three main allied jet fighters stationed in Korea at the time; the Navy's supreme fighter, the F9F Panther, the F2H Banshee, and the US Air Force's F-80 Shooting Star, which was just about outdated by 1950. So there was no hiding; the allies had to find a solution to the MiG menace; the Navy's Panthers and Banshees could not do it alone.

F-86

On November 7, George E. Stratemeyer, the Far East Air Forces commanding general, also had his problems with the MiG threat. He demanded that his fighter strength be supplemented by something more powerful than the F-80. The next day, he was guaranteed one wing of Republic F-84s and one wing of North American Aviation F-86As. The United States authorities ordered the loading of the aircraft in San Diego on the escort ship USS Bairoko (CVE-115) and the light carrier USS Bataan (CVL-29) on November 14.

Consequently, the first known aerial battle involving swept-wing jet aircraft took place over Korea on December 17, 1950. Although the Soviet-made MiG-15 was introduced to the Korean War in November, its speed and maneuverability caused significant problems for the United States B-29 bomber force and its F-80 escorts. For the Navy, things were just as bad as its frontline fighters, the F9F Panther and the F2H Banshee, could not outmaneuver the fast and agile Soviet jet.

After a long Pacific trip from San Diego, the F-86As arrived at Kimpo on December 15. Just two days later, on December 17, the first kill of a MiG-15 by an F-86 was registered by Lt. Col. Bruce Hinton, commander of the 336th Fighter Interceptor Squadron, 4th Fighter Interceptor Wing. Hinton led a flight of four F-86s over northwestern North Korea. To fool the Soviets, the USAF Sabres flew at the same height and speed as F-80s on sorties and used F-80 call signs. The Mig pilots were fooled into thinking they were on another turkey shoot. Hinton saw four MiGs at a lower altitude and led his aircraft in an assault. One of the MiGs went up in flames after being hit by a burst of machine gun fire from Hinson's F-86. The battle was clearly won by the new American jets. The tables were turned one month after the Soviets reclaimed the skies from the allies. It was primarily due to the F-86 jets. The MiGs would no longer rule the Korean skies.

The US Navy was not in a good position in Korea at the end of 1950. It realized that its Panthers and Banshees would not be able to dominate the nimble MiGs in dogfights. The F9F Panther was designed at the same time as the F-86, which was the first swept-wing jet fighter to be produced in the United States. The F9F and F-86 prototypes made their first flights within two months of one another, but they were a world apart. The swept-wing jet technology, developed by the Germans during World War II and taken by the Americans, was made accessible to the engineers of North American Aviation (NAA) and the engineers of Grumman. The USAF was very interested in the new F-86 jet being developed by NAA, with its 35° swept wings. At the same time, Grumman was more cautious and developed the straight wing F9F-Panther. Grumman being the preferred manufacturer for the Navy meant that the Navy would order the Panthers. The result was that the USAF could take on any threat with its fast and highly maneuverable F-86s, while the Panther had to fight battles with the MiG 15s that were beyond its capabilities.

In reaction to the MiG-15's unexpected emergence and clear superiority against the Navy's fighters, the Bureau of Aeronautics launched a search for a replacement for the straight-wing Panthers and Banshees in service. The sweptwing McDonnell F3H-1 Demon was already in development for the Navy, but the powerplant issues that would haunt and later doom that aircraft were already apparent in the early 1950s.

The Bureau's search revealed three suitable aircraft. However, none of them would be available to the Navy soon. By the time they were ready for service, the Korean War had ended. The first of three types that the Navy Bureau chose were the F-86-derived Fury planes from North American Aviation. They entered Navy service after the Korean War.

The second type was Grumman's improvement to the Panther, the more desirable F9F 'Cougar' with its swept wings and more powerful engine. The Cougar entered service before the end of the Korean War in November 1952 but did not participate in the conflict. The third type was Vought Corporation's F-8 Crusader. It was arguably the finest of the bunch. Unfortunately, the F-8 entered service quite late, in March 1957. However, the wait was worthwhile. It was one of the best fighters in the world when it went into service with the US Navy. It remained in service with the Navy for 19 years and another 15 years with the US Naval Reserve.

THE ATTACKS ON NORTH KOREA'S DAMS

Peace talks between the UN and Communist forces made little progress in early 1952. Many members of the US military were dissatisfied as a result of this. In May 1952, the new US Far East Commander, General Mark Clark, advocated the deployment of air power to compel the enemy into yielding to UN requests. He was not afraid to approach the Joint Chiefs of Staff (JCS), whose members were similarly fed up with the never-ending discussions.

The Far East Air Force (FEAF) decided on an aggressive campaign. The significant targets for the escalated air campaign would be the North Korean hydroelectric plants.

The operation began on the afternoon of June 23, and the attack on the Suiho dam is noteworthy as an example of interservice collaboration. It started with 35 Navy F9F Panther jets neutralizing enemy defenses, then 35 Navy AD Skyraiders with 5,000-pound bombs going in to bomb the Dam. All the aircraft involved in the operation were from Task Force 77 of the Seventh Fleet, which was operating four fast aircraft carriers for the first time. Ten minutes after the aircraft from Task Force 77 left the scene, 124 F-84s of the Fifth Air Force flew in and attacked the target, while 84 F-86s provided protection throughout the mission. In just four days, with 546 Navy sorties and 730 by the USAF with its fighter-bombers from the Fifth Air Force, 90 percent of North Korea's electric power capacity had been destroyed. Due to interservice rivalries, such combined aviation operations by the Navy and Air Force would have been unthinkable earlier in the War. After the 'Revolt of the Admirals,' the Navy and Air Force would not cooperate on any level, and each service was granted its zone of action. However, by 1952, the personal ties that General Clark and General Weyland of the Air Force formed with Vice Adm. Joseph 'Jocko' Clark of the US Navy's Seventh Fleet created an environment that promoted more collaboration between the two services. This friendlier environment enabled the attacks on the Dam to be completed without any hiccups.

The Suiho Dam on the Yalu River, in March 1953, almost one year after the attack by the US Navy and USAF. (USAF)

F9F PANTHER OPERATIONS DURING THE KOREAN CONFLICT 1950-1953

This section offers the reader a glimpse of the types of missions flown by the US Navy's fighters during the Korean War. It mainly contains details of F9F Panther missions; however, aircraft like the F4U Corsair and AD Skyraider are also mentioned.

1950
July 19

In a pre-dawn launch, CVG-5 sent 13 AD Skyraiders, 19 F4U Corsairs, and 24 F9F Panthers to the enemy regions in the North. The target was the Bongun chemical plant at Hamhung, with secondary importance given to the Kogen railroad bridge.

The F9F targets in the region were Yonpo, Sondok, and West Airfields. The F9Fs found no aircraft at Kanko West and flew to the other two airfields. At Yonpo, there were 12 planes on the ground and 15 at Sondok, which were Yaks and Il-10s. The F9Fs flew 20-25 concentrated runs on the two airfields and destroyed 13 aircraft, seven at Sondok and six at Yonpo, while another six aircraft were somewhat damaged. Ten attacks were also made on the airfield hangars, which received heavy damage but were not ablaze. On the flight back to the carrier, the F9Fs staffed a power substation in the east, which sustained some damage.

The most spectacular destruction occurred at Inchon. In that city, a group of F9Fs headed for a fuel storage installation. The Panthers went to work on the plant with their 20mm cannon and destroyed seven colossal oil storage tanks and two smaller ones, which blazed wildly, causing towering flames and black smoke to extend thousands of feet into the sky. Moreover, leaking fuel seeped throughout the principal oil storage facility and caught fire, which resulted in further damage.

July 25

The mission for the day was to stop the advance of North Korean ground forces in the critical southwestern sector of Korea.

At 9:45 am, 8 F9F Panthers were launched, with their mission being to sweep the target area. Unfortunately, one Panther suffered mechanical difficulties and did not join the others, so only seven jets made the offensive sorties.

Returning pilots reported little enemy advances. The only troops seen were twenty to twenty-five on a bridge midway between Songjong-ni and Pochon-ni. Strafing from the F9Fs seemingly killed six to ten of the enemy. There were no other North Korean People's Army (KPA) soldiers in the area, only farmers working the fields with their families. With little evidence of further military movements, the F9Fs were on their way back to CV-45 when some trucks were spotted. The Panthers destroyed one truck, with three others receiving some damage.

A large power plant was sighted and attacked by the F9Fs with 3,500 lb. bombs. One bomb hit the target and destroyed about a quarter of the facility. Then further down, a railroad tunnel was bombed and damaged in both entrances, rendering the tunnel useless.

The second F9F sweep from USS Valley Forge consisted of eight offensive sorties. Amazingly, no enemy military activity was noted along the lines of communication. Two parked command cars at Kwangju Airfield were damaged, while a locomotive with camouflaged boxcars hit in a previous sortie was strafed again and destroyed.

Two trucks parked near houses ten miles west of Kwangju were sighted and were severely damaged by the Panthers.

July 26

Eight offensive softies took off at 09:45, with two divisions of F9Fs taking to the air. They struck communication lines between Suwon and Taejon. There were few other threatening targets in the area where the F9Fs flew over. Near Chonan, the Panther pilots noticed a string of well-strafed boxcars between Suwon and Taejon. The road was littered with hundreds of destroyed vehicles. Further up, at Chochiwon, one division leader strafed the only truck which appeared intact, damaging it. Near the truck, the pilot saw around eight destroyed tanks.

August and September

Operations against North Korean invasion forces were conducted by CVG-5 on August 26, 27, 29, 30 August, and September 1, 2, and 3.
During this period, CVG-5 operated from the USS Valley Forge (CV-45) in company with the USS Philippine Sea (CV-47) and CVG-11. Flight operations against the North Korean forces comprised controlled close support work, armed reconnaissance patrols, sweeps, and interdiction-strike work on assigned targets by small and medium size strike groups. Among the aircraft used on this critical operation were two F9F-3 squadrons of 28 Panthers, with 39 pilots available. The CVG-5 group on Valley Forge provided close support to the front-line ground forces on August 26 and September 1, 2, and 3. Bad weather prevented similarly scheduled action on September 5. Routine CAP, ASF, or utility flights were also conducted along with offensive operations during the following dates; August 28 and 31 and September 5 and 6.
After the end of operations, one F4U-4B ditched at sea after being hit by North Korean AA fire. There were also 15 other aircraft that received battle damage from enemy AA or friendly fire. Two aircraft required transfer for heavy repairs, while five others were repaired aboard the carrier. Damage to the remaining eight aircraft was minimal. As a result, there were no casualties on the US side. In contrast, six enemy aircraft were destroyed, and another five were damaged.

1951
August 25

CVG-5s F9F-2 Panther and F2H-2 Banshee jet aircraft, off USS Valley Forge, were used exclusively on armed reconnaissance flights and day Combat Air Patrols (CAP). On August 25, 23, jet aircraft escorted USAF heavy bombers on a high-altitude bombing mission. It was the first time that naval jet aircraft escorted US Air Force planes over enemy territory during the Korean Campaign.

October 28

On October 28, 1951, a special strike mission was flown against a vital supply center near Kapsan, North Korea. The strike was led by CVG-5, off USS Essex (CV-9), and was composed of 4 F2Hs, 8 F4Us, 8 ADs of CVG-5, and 4 F9F, 8 F4U, and 8 F4D aircraft of CVG-15. The jets preceded the strike group in the attack, strafing and rocketing known AA positions. They were followed by the Corsairs armed with VT bombs for flak suppression. Finally, the AD Skyraiders, armed with 1,000 lb. and 250 lb. napalm and incendiary bombs, followed the Corsairs and bombed specifically assigned targets of warehouses and barracks. The evaluation of post-strike photographs showed that 35% of the target was completely destroyed, which was considered excellent for the number of aircraft involved. The tactics used were basically those of the second world war. Although intense AA fire was encountered, no plane was shot down, even though six runs were made. After each run, a complete rendezvous of the strike group was made, and the target was approached from a different direction each time.

October

The F9F-2 squadron on USS Valley Forge used four plane reconnaissance flights successfully. In some instances, the planes separated during flight, but pilots became accustomed to keeping the other aircraft in sight after a few flights. Several jets were diverted from strike or reconnaissance missions to fly 'Target Combat Air Patrol' (TARCAP) when enemy MiGs appeared in the area. At that time, the Navy pilots found it difficult to differentiate between the F-86 Sabres and the MiG-15s. In order to avoid friendly fire incidents, all Navy pilots were briefed on recognizing the differences between the US F-86 and the Soviet MiG-15. Furthermore, on all reconnaissance flights, the doctrine was for the second section to be used as aircraft lookouts.

1952
February 2

LTJG L. R. Cheshire of VF-51, from CVG5, on the USS Essex (CV-9), flying an F9F, was killed in Wonsan Bay. His F9F Panther was struck by AA on his second run attack on a camouflaged North Korean train, and he headed for Wonsan Bay to ditch the aircraft. His jet was on fire, but he was on the verge of making a perfect splashdown near a destroyer when his ejection seat discharged from the plane, which was practically on the water. The destroyer arrived quickly, and despite a thorough search of the area, the destroyer's search crew could not locate LTJG Cheshire.

February 19

During the Korean War, the United Nations launched a determined interdiction operation on North Korea's supplies and communication, which began in August 1951. USS Valley Forge's F9Fs were heavily involved in the operations. For instance, on February 19, 1952, 68 sorties were flown from the carrier in support of 'Operation Strangle.' The attacks by the F9Fs scored 75 rail cuts and 121 enemy troop kills. They also completed the destruction of 21 buildings, 15 oxcarts, and four railroad bridges and inflicted other minor damage.

March 17

On March 17, six US Navy F9F aircraft suffered flak damage during combat missions. One of the six had to land in South Korean territory at King 18 airstrip. The other five made it aboard USS Valley Forge. For the day, a total of 90 sorties were flown, scoring 88 rail cuts.

June 23 – June 25

The following section further details the activities carried out by the carrier USS Bon Homme Richard (CV-31) during the dam assaults on North Korea, as indicated in the preceding section.

USS Bon Homme Richard accommodated CVG-7 in June 1952. CVG-7 had three F9F squadrons; VF-71 and VF-72. Each squadron had 16 F9F-2s, and the photo-reconnaissance squadron, VC-61, with 3 F9F-2Ps.

A modified strike timetable was issued as part of a concerted attempt to destroy North Korean power facilities. The assaults were carried out by planes from four carriers: the USS Boxer, the USS Bon Homme Richard, the USS Philippine Sea, and the USS Princeton. Several hundred US Air Force jets were also involved to ensure maximum damage to the selected targets.

On June 23, Lieutenant Commander (LCDR) A.N. Curtis, the Commanding Officer (CO) of VF-72, with 11 F9F-2s, struck Kyosen's No.2 hydroelectric plant, about 30 miles Northeast of Hamhung. The transformer yard was severely damaged, and a direct hit on a gasoline or oil storage tank started a large fire. CO J. S. Hill of VF-71, with 7 F9F-2s, hit Fuson's No. 2 hydroelectric plant west of Kyosen. VF-71 got two direct hits on the generator building, caving the roof in, and starting a fire. In addition, severe damage was dealt to the transformer yard and surrounding structures. This strike put Fuson's No. 2 hydroelectric plant out of commission.

There was no flak at the Kyosen No. 2 facility and only minimal, inaccurate flak at the Fuson No. 2 plant. One AD Skyraider, however, was severely damaged by flying concrete after the lead plane bombed the generator facility, causing big cement fragments to strike the AD. The pilot was extremely fortunate to survive, thanks to a newly installed armor plate on the AD. That was one of the reasons why the Skyraider was regarded as one of the toughest prop bombers in Naval and Air Force aviation history. Furthermore, despite being a single-engine aircraft, it could carry as much ordnance as a Boeing B-17 Flying Fortress over a comparable range when properly equipped. By the way, the Skywarrior pilot safely landed on his allocated carrier. I digress; back to the F9Fs.

The aircraft were launched at 8:00 am on June 24 to strike the North Korean hydroelectric complex. This time, All CVG-7 aircraft were sent to destroy the Kyosen No. 4 facility, six miles inland from Tanchon. Complete destruction of the entire target area was achieved. No flak was encountered due to the flak suppression runs made before the strike by VF-74's Corsairs. Following the strike on Kyosen's No. 4 plant, VC-61s Photo Recon F9F-2P Panthers took photos of the entire Kyosen complex, confirming the damage.

Skyraiders, Corsairs, and Panthers attacked railways and bridges northeast and northwest of Hungnam in the afternoon. Four railroad bridges were severely damaged, and fourteen rail cuts were made. Again, no flak was experienced during the attacks.

Because of poor weather on the morning of June 25, the assaults were rescheduled for the afternoon. Skyraiders, Corsairs, and Panthers took off at 12:30 pm to strike targets around Wonsan. Several targets were hit successfully.

Forty-four offensive sorties were launched from the carrier in the three days of operations. Then, on June 26, the USS Bon Homme Richard replenished at sea on the way to Sasebo Naval Base in Japan. Finally, on June 27, it arrived, where it resumed ready carrier status.

Regarding the two F9F-2 squadrons aboard the USS Bon Homme Richard, VF-71 was disestablished in 1959, and VF-72 'Bearcats/Hawks' was redesignated VA-72 in 1956.

June 27

CVG-11 used its F9F-2, AD Skyraider, and F4U Corsair aircraft to carry out coordinated group strikes. This time, the attacks were concentrated against troop and supply concentrations instead of railways. The early morning strike targeted a large troop concentration, a vast army stationing location, and a supplies storage site. In the afternoon, the aircraft targeted more supplies, truck parking areas, repair stations, and a troop concentration site. The total number of sorties flown was 119. Total ammunition used was 10,500 20 MM cannon rounds and 21,100 (50 Cal.) rockets. In addition, a total of 16 bombs and 5.2 tons of incendiaries were dropped.

November 1

On the morning of a terrible day, CVG-101 (CVG-101 was redesignated CVG-14 on February 4, 1953) operated heckler flights ('Hecklers' are small propeller-driven aircraft engaged in nuisance bombing) which covered main supply routes West and North of Wonsan. They managed to damage one locomotive and several rail cars. One coordinated strike was flown in support of the frontline troops employing AD Skyraider, F4U Corsair, and F9F Panther aircraft. Two pilots were lost on that bleak day. The first was LT C. O. Glisson, an F9F pilot of VF-721. He reported a rough-running engine and turned toward the east coast of Korea to head back to USS Kearsarge (CV-33). Unfortunately, he did not make it. He was last seen entering a heavily clouded area. Soon afterward, his aircraft crashed into the sea, about ten miles from the shoreline. He was pronounced KIA. The second pilot, LT R. G. RIDER, a VF-884 pilot of an F4U, was hit by AA fire while diving on a target in the vicinity of Chun-Chon. He did not recover from the dive and was KIA. Only 42 sorties were flown on that day due to heavy seas and overcast weather conditions,

November 18

On November 18, three CVG-12, VF-121 squadron F9F-5s were confronted by four MiG-15s. LT Royce Williams and LTJG John Middleton were each credited with one MIG-15 kill and LTJG David Rowlands with damaging one. The F9F pilots used sound defensive tactics and an alert lookout doctrine, which was a significant reason for defeating the enemy. Despite the MiG-15 's obvious performance advantage over the F9F Panther, the aggressive US Navy pilots pushed their aircraft to the maximum of their combat potential. By doing so, they took full advantage of the mistakes made by enemy pilots. In addition, they demonstrated the superiority of well-trained pilots compared to the lesser-trained opponents with better equipment.

In another operation by CVG-12, F9F-5 panthers were used as bombers. In fact, the F9F-5s were used as bombers on many occasions, and they proved to be a potent weapon, considering they could each carry as much as 2,000 pounds of bombs and use the Mk-8 catapult to launch the projectiles. For example, four F9F-5s of the VF-122 squadron, carrying two 1,000-pound bombs, attacked the heavily defended

Hamhung Bridge. The Panthers scored six hits out of a total of eight bombs dropped and damaged the bridge; however, they did not destroy it.

1953
January 16

Hecklers successfully targeted a temporary military camp south of Hamhung early in the morning, resulting in secondary explosions. AD Skyraiders and Corsairs worked together to attack supply shelters southwest of Songjin and a mining facility south of Kilchu. Eighteen buildings were destroyed, while another twelve were damaged. After flying north central reconnaissance flights, the F-9F-5s assaulted Simpo fishing sites. The strike results could not be confirmed due to poor visibility. In the afternoon, the Panthers flew flak suppression for the props, which were attacking supplies and troop built-up areas south of Changyon-Ni, near the eastern front lines. Two buildings were destroyed, and a subsequent explosion damaged eight. The F9Fs targeted railroad infrastructure near Yang-Dok, five boxcars were destroyed, and ten were damaged. That day, 52 tons of bombs were dropped from 85 sorties. The following day, Panthers attacked supply dumps just behind the enemy's front lines. Ten supply buildings were destroyed, and four were damaged.

March 6

On this day, prop aircraft attacked personnel and supply shelter concentrations east of Hamsong, with F9F Panther jets providing flak suppression. The area was 50% covered, but the damage could not be assessed. The Corsairs attacked coastal guns south of Wonsan, while AD Skyraiders provided close air support for the ground troops. One large gun, seven bunkers, and 150 yards of trenches were damaged. Furthermore, one large fire and four secondary explosions were seen. The F9F-5s, flying close air support for the first time, were credited with ten bunkers destroyed. Other Panther missions heavily damaged a power relay station near Tanchon, and four barracks buildings south of Wonsan.

All in all, 91 sorties were flown, and 73 tons of bombs were dropped before a flight deck accident forced the cancellation of air operations. A bomb hung under the wing of a returning F4U Corsair broke loose on recovery and exploded on the flight deck of USS Oriskany. Fourteen personnel were wounded, two others were killed; Aviation Electrician Airman Thomas M. Yeager, who was working on an F9F at the time, and the US Navcameraman, Airman Thomas Leo McGraw Jr., who tragically filmed his own death.

April 21

April 21 was an excellent day for the pilots. It was 'Boy-san Day' when the pilots could select their own targets. Seventy-one tons of bombs were dropped during the day's 109 sorties. In the morning, the first jet strike attacked billeting and supply buildings. Post-strike photos confirmed that 21 buildings were destroyed and six damaged. The second F9F Panther jet target was against a storage area northwest of Pukchong, which destroyed two warehouses and damaged two others.

In the afternoon, the F9Fs attacked a troop billeting area near Iwon and bombed the vital Hamhung highway bridge. The bombs hit three parts of the bridge. The first was in the middle section, which was demolished, followed by a second strike on a span of the bridge near the southern terminus. Finally, a span near the northern approach was bombed and heavily damaged in the third hit.

But that was not all; later, in another strike, the F9F Panthers managed to destroy or damage up to 50% of a truck camouflage area consisting of 177 buildings.

THE CAREFUL PILOT

A VF-61 squadron pilot in an F9F-2 in 1951 accomplished one of the trickiest carrier landings in naval aviation history. Lt. (JG) John P Eells brought his Panther in for an emergency landing on the Franklin D. Roosevelt (CVB), the second of three Midway-class aircraft carriers, without causing significant damage to the Panther. While preparing to land, Lt. Eells noticed that his F9F's left main gear failed to extend. Knowing he was facing a very tricky situation with only a nose wheel and the right main landing gear, he tried to 'unstick' the left wheel. After all possible means of lowering it were exhausted, he warned the carrier of his predicament and started discharging excess fuel. After that was accomplished, he entered the pattern, executed a perfect pass, and the Grumman 'niner' was now coming in to land. For the crew on the carrier, it was like watching a slow-motion movie. The aircraft was at minimal speed and flying in cautiously. Tension was everywhere on deck. In the 1950s, many bad accidents occurred on carriers, even with aircraft that had no problems until the last moments of landing, such as the tragic crash of Lt. Cmdr. Jay Alkire in an F7U-3 Cutlass, on the USS Hancock (CVA-19) on July 14, 1955. After catching the wire, Eells's plane miraculously rolled forward on only the nose and right main gears and came to a stop. At this time, the aircraft was in equilibrium before the F9F's left wing dropped, with the left wing tip hitting the surface, causing only minor damage to the tip tank. After a change of tank for the left wing, the repairs of the left wheel well, and the door timing mechanism, which were the cause of the wheel door not opening, the Panther was all right to go within a few days. More importantly, there were no casualties. Surely, Lt. (JG) John P Eells deserved something for his unique landing.

An experimental F9F Panther on board the USS Coral Sea CVA-43 for Carrier Qualifications in 1952. (US Navy)

F9F-5 Panther, (BuNo 126204), in flight over mountainous Korean terrain, on June 14, 1953. This Panther is from Fighter Squadron 111 'Sundowners' (VF-111), based aboard USS Boxer. VF-111 was disestablished in 1995. (Naval History and Heritage Command)

A Marine F9F-2, (BuNo 123440), VMF-311, redesignated VMA-311 in 1957, at Hamilton AFB, California, in 1950. This aircraft stalled and crashed while trying to land on the USS Philippine Sea (CV-47) off the Korean coast, on February 16, 1952. The pilot was killed. (Bill Larkins)

A Marines F9F-5P, BuNo 125320, in Korea. This model was the photo recon version of the F9F-5. Thirty six of these were built. The camera window is visible on the front left of the aircraft. (National Archives)

This Grumman F9F-2 Panther (BuNo 123494) of fighter squadron VF-21 'Freelancers' went through the barrier and crashed into three other jets on the USS Midway (CV-41), which it carried with it into the sea by the collision. The pilot was safe and sound, not so for his bruised ego. He was picked up by helicopter. (US Navy photo)

An F9F-2 Panther, BuNo 123018, of NATC in flight over some excellent terrain around Naval Air Station (NAS) Patuxent River, Maryland, in the 1950s. (USN photo - Commander Richard Timm, Naval History and Heritage Command)

This F9F-2B is from Fighter Squadron VF-721 'Iron Angels', from the carrier USS Boxer (CV-21). Seen here on an armed recon mission over North Korea, in July 1951. VF-721 was redesignated VF-141 in 1953. (USN photo National Naval Aviation Museum)

A Marines F9F Panther, fully loaded with bombs, takes off from an airfield in an arid region of Korea, during the Korean war. (National Archives)

An F9F-5 Panther belonging to Fighter Squadron VF-742 'Ironmen', of the Naval Air Reserve, on the deck of the carrier Antietam (CVA 36), which was operating off Guantanamo Bay, Cuba, on February 9, 1953. In 1953, VF-742 was redesignated VF-82. (US Navy Photo, Collection of National Naval Aviation Museum)

F9F-2 Panther of Carrier Air Group Seven (CVG-7), with Fighter Squadron VF-72 'Bearcats/Hawks', is launching from the deck of the carrier USS Bon Homme Richard (CV-31), for a strike against North Korea, on June 23, 1952. In 1953, this Panther joined squadron VF-151 'Vigilantes'. (US Navy Photo, Collection of National Naval Aviation Museum)

The story of the 'Blue Tail Fly' F9F-5, above, is interesting. One pilot from USS Princeton, Lt. (jg) Richard Clinite of VF-153, was on a mission on May 5, 1953, when his Panther was struck in the tail region by flak on May 5, resulting in extensive damage. Clinite's F9F was not in the all blue of other Panthers at the time, but in the new experimental unpainted aluminum finish, which was applied to about 100 Panthers in 1952 by the Navy. The Navy took the idea from the USAF which had most of its jets in the natural metal finish in the 1950s. The new style was better economically, as the aircraft in the natural metal finish cost the Navy less in operational expenses, and a decrease in the weight of the aircraft.

The Panther was planned to remain grounded until substantial repairs could be completed. However, that was not the case. In a second incident, that occurred shortly after Clinite's ordeal, Ensign W.A. Wilds returned from a flight with substantial flak damage to the forward fuselage of his Panther. In this case, his Panther was painted in the standard all blue color. Some wise person in the hangars of USS Princeton, had an idea, why not take Clinite's front all metal section, which was undamaged, and merge it with Wilds' blue tail section, which was also in good condition? The maintenance crew got to work and completed the task in no time. The result was the 'Blue Tail Fly'. Clinite was back in his all metal Panther, minus the tail section which was blue. The 'Blue Tail Fly' was flown on several missions until flak damaged it to the point where the F9F needed major repairs done.

Tragically, for Lt. (jg) Richard Clinite, his fate was preordained. The first incident, when his tail was blown off, did not get him, but the next one did. A short eight days after his F9F was hit by flak and grounded, he took off on another combat mission with an F9F Panther, on May 13, 1953. During combat, the Panther Clinite was flying sustained a significant flak hit, forcing him to eject over the sea. Tragically, Clinite drowned because the wind prevented his parachute from collapsing, and the rescue helicopter was unable to carry him to safety due to his chute ballooning. It was a tragic end, for the daring pilot. (US Navy Photo, Collection of National Naval Aviation Museum)

F9F-5 Panther of Fighter Squadron VF-72 'Bearcats/Hawks' plows into the barricade of the carrier Tarawa (CV-40) during flight operations, on August 19, 1952. This aircraft was previously with the 'Blue Angels' Aerobatic Team. Squadron VF-72 was redesignated VA-72 on January 3, 1956. (US Navy Photo, Collection of National Naval Aviation Museum)

This is the photo reconnaissance version of the F9F-5, of which 36 were built. This one was modified to a F9F-5KD drone director aircraft, and redesignated DF-9E in 1962. It was with the VU-1 Fleet Utility Squadron. Photo taken in 1970. (RuthAS)

This F9F Panther is showing off its six underwing rockets. Photo taken on August 2, 1950.
(Naval History and Heritage Command)

F9F-2B Panther of Fighter Squadron VF-191 'Satan's Kittens' off the carrier Princeton (CV-37) in flight during the Korean War. Squadron VF-191 was disestablished in 1978, and re-established in 1986. However it was disestablished for good in 1988. (US Navy Photo, Collection of National Naval Aviation Museum)

THE EYES OF THE FLEET

VC-61 - THE NAVY'S PHOTO RECONNAISSANCE SQUADRON

VC-61 flew the Navy's first combat photo mission with jet aircraft from USS Princeton in December 1950. The unit went into combat during the 'Battle of Chosin Reservoir' evacuation of the Marines. The Battle of Chosin Reservoir started a month after the People's Republic of China entered the War on November 27, 1950. In a surprise attack, 100,000 Chinese troops surrounded American forces in some of the region's roughest and most inaccessible terrain, where temperatures often dropped to 25 degrees below zero. American soldiers were besieged, massively outnumbered, and faced mass slaughter. Finally, they had to retreat from the 'Icy Hell' in the northern regions of North Korea.

The successful bridge and tunnel interdiction program against the enemy was primarily made possible through photos taken by the F9F-2Ps of VC-61. More than 500 bridges were photographed, some five or six times. As soon as the bridges were repaired, the photos confirmed it, and new strikes were launched against them.

Part of the interdiction program was surface bombardment from American battleships, cruisers, and destroyers. Low oblique photos were needed in that situation, like a coxswain's view of the area. The F9F-2P jets delivered pictures to the ships to locate targets and pinpoint their gunfire to the Chosin reservoir to aid the Marine withdrawal. After a 17-day battle, the weather worsened, and the US forces made it to relative safety in a fighting retreat. Unit CV-61 played its part in the 'Chosin Reservoir' operation very well.

While in Korea, the unit operated over the area from Chongjin to Wonsan and North to the Yalu. Lt. C. A. Cooper was the officer-in-charge of the first Navy jets to be used for aerial photographic work. The unit started with three pilots, Lt. C. A. Cooper and Lts. (J G) John E. Smith, and George Elmies, There was also a crew of 15 enlisted men in the unit. VC-61 had three F9F-2P Panther jets for its three pilots.

The unit's photo equipment was mounted in the nose section of the three Panthers. The F9F-2P's front area easily accommodated both the vertical and oblique cameras. NAS Alameda's Overhaul and Repair (O&R) workshop had removed the Panthers' 20 mm cannon to install the camera equipment. However, there was a slight problem; the cameras were lighter than the removed weaponry. Six hundred pounds lighter, that is. So Almeda's O&R added 600 pounds of ballast to the front

Marines halted on the road between Yudam-ni and Hagaru-ri, while other Marines are clearing the way for further advance. It was an 'Icy Hell' on earth. This photo was taken in the middle of the Battle of Chosin Reservoir, on December 1, 1950. (Oliver P. Smith Collection (COLL/213), Marine Corps Archives & Special Collections)

Sailors working with a tow bar attached to an F9F-2P Panther of Composite Squadron VC-61, Detachment A on the flight deck of the carrier USS Boxer (CV-21) operating off Korea on May 14, 1952. (US Navy Photo, Collection of National Naval Aviation Museum)

of the aircraft to make up the weight differential. That did the trick. Now, all VC-61 had to do was take photos.

VC-61's pilots liked the Panthers for their stability and better visibility for both the cameras and pilots. They also had excellent range; their speed was 18 mph, faster than the standard F9Fs. Furthermore, oil did not smear the photo window on the photo-reconnaissance Panthers compared to the prop aircraft, which carried their cameras in their belly region, which was very oily.

Jet photo planes could shoot pictures at all altitudes from 100 feet to 40,000 feet and at speeds ranging from 175 mph to 575 mph. Regarding the equipment, the unit used Sonne cameras, which were Leica copies made by the Italian manufacturer Antonio Gatto in 1948, and were used by VC-61 for low-altitude reconnaissance photos. In addition, K-17 aerial cameras with 6- to 12-inch focal lengths were used for vertical work and 6 inches for horizontal and oblique exposures.

Jets are usually considered high-speed aircraft, but photo pilots would slow them down and fly low on photo assignments. Most of VC-61's missions were at altitudes under 5,000 feet, with speeds ranging from 175 to 230 mph. The planes performed well at low power settings, and fuel consumption was economical. The average photo hop was about two hours long. During that time, the F9F-2Ps covered as much as 600 miles. Conventional planes remained in the air for about twice as long but seldom covered more than half the distance of the photo aircraft. Photo pilots used their speed and altitude going to and from the target and dropping down, working low and slow at their assigned target area.

Lt. Zack Taylor gets ready for a recon flight over enemy territory in his F9F-2P, while the carrier was operating off Korea in April 1952. Note the camera window above 77, and Lt. Taylor's ribbed crash helmet.
(Naval History and Heritage Command)

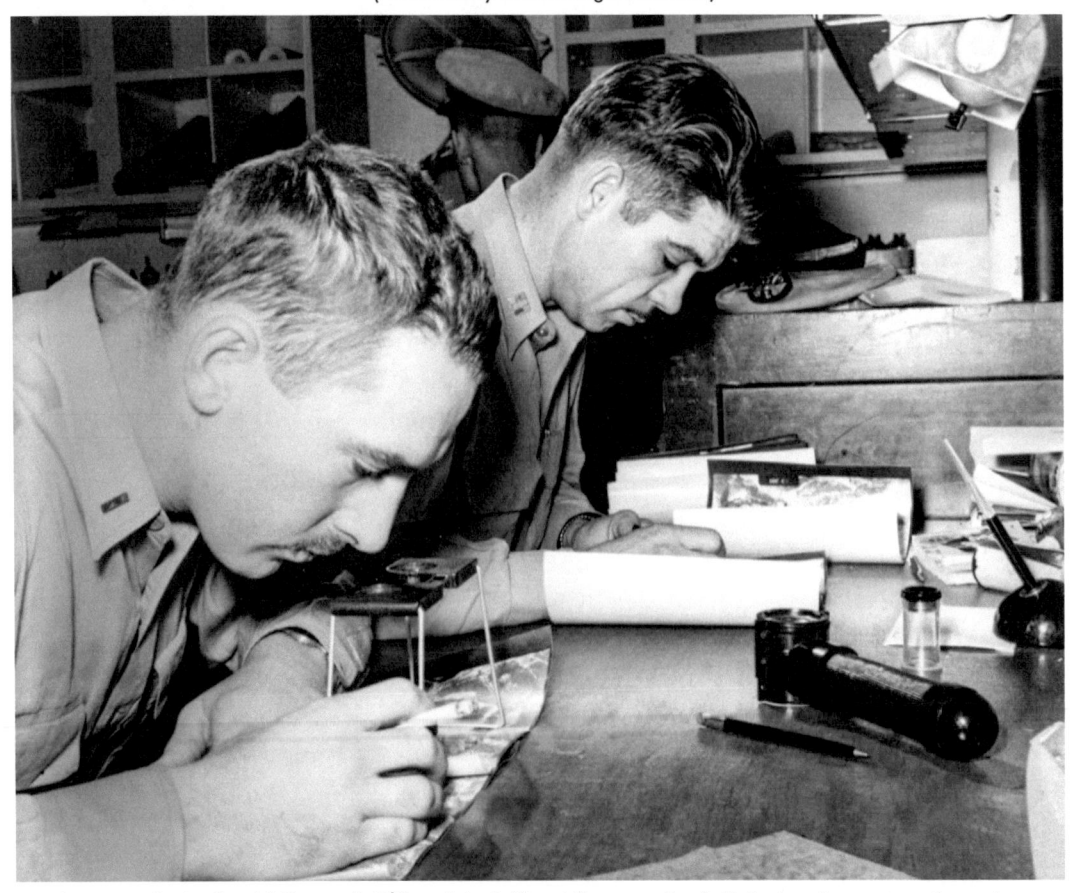

Lieutenant (Junior Grade) George B. O'Connel, Jr., (left) and Lieutenant Roy L. Hall, photo interpreters of squadron VC-61, examine reconnaissance photography of enemy concentrations. (Naval History and Heritage Command)

Each photo recon plane was accompanied on missions by one jet fighter, which provided protection from enemy fighters, or if the photo aircraft was hit by ground fire. The photo pilots had no guns and had to use their speed and the help of their fighter escort to avoid getting hit. Enemy ground fire was a significant menace to all aircraft in the Korean War, due to the installation of modern anti-aircraft batteries in all critical areas, such as bridges, structures, factories, etc., by the North Koreans. The Navy's photo planes had to work a great deal on flak coverage. Photo interpreters studied their pictures to locate flak installations, which was quite tricky.

The jet photo aircraft were easy for the deck naval officers to fix. The sliding nose permitted easy access to all cameras and equipment. The F9F-2Ps were very reliable jets and were always up for the task. There was only one occasion on the USS Princeton when a photo mission had to be canceled. Not a bad statistic for the reliable Grumman Panther jets and their pilots.

After six months on the carrier USS Princeton photographing North Korea in 160 combat sorties, VC-61 was ready to return to the US for a well-earned break. Before completing its six months tour of duty, VC-61 had taken 14,000 pictures of North Korea showing cities, bridges, troop concentrations, airfields, gun emplacements, and supply dumps.

In addition, the unit mapped countless sections of Korea, including the east coast from the 7th parallel to the Russian border. Their work was an invaluable source of intelligence information for the entire US Navy fleet.

This is one of the three F9F-5P Panthers of VC-61, BuNo 126270. In 1956, the aircraft was transferred to the Marines squadron VMCJ-3. VMCJ-3 was commissioned on December 12, 1955, after the merger of VMC-3 and VMJ-3, which combined the electronic warfare, and photo reconnaissance squadrons of the Third Marine Aircraft Wing into a single squadron.
(US Navy Photo, Collection of National Naval Aviation Museum)

THE BLUE ANGELS

In 1946, after the Second World War, the euphoria of victory was subsiding. The Naval aviators, who bravely fought far away from their loved ones in the War, started leaving the Navy for greener pastures. They were getting married to their sweethearts, raising families, starting businesses, applying for better-paying jobs, or going to university to get a higher education. As a result, the Navy was losing good men by the thousands, and no recruits were knocking on the Navy's doors.

On April 24, 1946, the famed Admiral of the fleet, Chester Nimitz, sent a brief note to Rear Admiral Ralph Davison, head of the Naval Air Advanced Training Command (NAATC). The message was for a request to Vice Admiral Frank Wagner, the Chief of Naval Air Training (NAT), located in Pensacola, Florida, to establish an aerobatic squad that would be linked to the command. Nimitz wanted to set up the aerobatic team to increase the American people's interest in naval aviation and, at the same time, raise the Navy's morale. The primary objective, however, was to assist the Navy in garnering public and political support for a higher portion of the ever-shrinking military budget.

Rear Admiral Ralph Davison personally chose Roy Marlin Voris as Officer-in-Charge and Flying Leader to assemble and train the new flight demonstration squadron. The choice of Voris, or 'Butch' as he was known, was a good one. He was a Second World War Ace flying F6F Hellcats in the Pacific theater, with seven kills to his name. In addition, the dogfighting skills he gained during the War served him well in his postwar career as the commander of the 'Blue Angels.'

In 1946, the Blue Angels chose their first aircraft, the esteemed Grumman F6F-5 Hellcat, Voris' aircraft, during the War. The pilots trained twice a day for just under a month and were ready for their first flying display on May 10, followed by the first public demonstration at Graig Field in Jacksonville on June 15. The performance lasted only 12 minutes, yet it had a lasting effect on the audience. Only three planes were used in the display, with the fourth kept in reserve. In that first show, the pilots were Roy 'Butch' Voris, Lt. Maurice 'Wick' Wickendoll, and Lt. Mel Cassidy; all three were Pacific War veterans.

Yet the team did not have a name. An officer from the headquarters of the Navy proposed the term 'Navy Blue Lancers,' but it was rejected by the team members. Finally, Lieutenant 'Wick' Wickendoll suggested the name 'Blue Angels' after seeing an advertisement for the nightclub 'Blue Angel' in a New York magazine. The rest of the squad adopted the name, and it became official. Thus, the first 'Blue Angels' demonstration flights took place in Omaha from July 19 to 21, 1946.

Later that year, the team switched to the Grumman F8F-1 Bearcat. The F8F was a beast. It was 20% lighter than the Hellcat, which resulted in a climb rate about 30% greater than the F6F. If the Bearcat had fought in World War Two, it would have performed exceptionally well due to its immense power (it was powered by the Pratt & Whitney 2100 hp R-2800-34W engine, with 2,000 horsepower (1,500 kW)) and maneuverability. After three years of flying the Bearcat, in 1949, the 'Blue Angels' decided to change to the new jets, so they swapped the F8F for the Grumman F9F-2 Panther. On August 20, 1949, in Beaumont, Texas, the Blue Angels gave their first public demonstration while flying in their brand-new F9F-2 Panther jets. The show was a huge success.

Unfortunately, in 1950, the Navy disbanded the Blue Angels in reaction to the Korean War, and the crew reported to Fighter Squadron 191 (VF-191), 'Satan's Kittens,' aboard the aircraft carrier USS Princeton. Unfortunately, the 'Blue Angels' lost one of their own in the War. Lt. Cmdr. Johnny Magda was shot down while flying a mission close to

Wonsan in March 1951 while operating off the north coast of North Korea. The premature death of Magda was the first 'Blue Angel' casualty in combat.

In 1951, The 'Blue Angels' were back in action, and they received some new equipment. The more formidable F9F-5 variant. The '-5' version was the definitive straight wing Panther model, equipped with the more powerful Pratt & Whitney J48-P-4/6A with water injection, 6.250 lb. s.t. (2.835kg), 7,000 lb. ab. (3,175kg), which gave it a top speed of 579 mph, and a climb rate of 5,090 ft/min.

The crew relocated from NAS Corpus Christi to NAS Pensacola, Florida, in June 1955 and spent the following winter there for winter training. That year, the team also received their first swept wing aircraft, the F9F-8 Cougar. That version replaced the '-5' variant in December 1955. They lasted until 1957 when the F-11F-1 Tigers replaced them. The final performance in the F9F-8 Cougar in the 'Blue Angels' colors was over Naval Air Station Pensacola on July 12, 1957.

The 'Blue Angels' demonstrate one of their formations on December 8, 1952, with their Grumman F9F-5 Panthers. (Naval History and Heritage Command)

The swept-wing successor to the Panther, the F9F-8 Cougar was with the Blue Angels for only a brief time, from 1955 to 1957. (US Navy Photo, Collection of National Naval Aviation Museum)

THE F9F PANTHER IN THE KOREAN WAR

No one could say that the F9F Panther, the Navy's first fully jet-powered aircraft, was the best fighter in the Korean War. It was slower than the Air Force's gleaming silver F-86 Sabres, with their swept wings and colorful schemes. However, the first 'jet versus jet' kill in Korea was achieved by the rugged, sturdy, and reliable F9F Panther. The Navy's jet was capable of significant feats with its four 20 mm cannon in its nose section, six 5-inch rockets, and two 500 lb. bombs under its wings. The Panther just did the job and did it without fuss. A MiG-15 wouldn't want to meet one down a dark alley on its own, MiG alley, that is. That was how the Panther was - a feline with claws ready to slash the opposition with vicious tenacity. The Navy's Panthers conducted more than 78,000 flights against the enemy throughout the protracted and bloody air campaigns over Korea, with 64 Panther losses to enemy action. The Panther played a more noteworthy part in the Korean conflict than many other aircraft and took its place on the US Navy's Pantheon of 'Immortal fighters.'

With the end of the Korean War, it was time up for the Panther as well. Many Panthers managed to continue to see service with the Naval Air Reserve, where they provided several pilots with their initial experience in jet-powered combat aircraft. Finally, it was time for the new kid on the block to announce himself to the world. It was the turn of the slick Grumman F9F Cougar with its swept-back wings and superior speed to make its mark.

GRUMMAN F9F-6 COUGAR

XF9F-6

Not long after the BuAer ordered two 'Panther' prototypes, the XF9F-2 and XF9F-3, the Navy Bureau asked Grumman if it would consider developing a swept-wing variant of the 'Panther.' Although Leroy Grumman, vice president Bill Schwendler, and the design team had begun a swept-wing study during the War in 1945, they declined to submit the Navy's requested proposal for a swept-wing jet. Long Island's technical team believed that much more research was required before the data at their disposal could be successfully applied to a project. Although disappointed, the BuAer agreed to wait until much more data was available.

With the F9F-5 Panther, a capable 630 mph carrier fighter that introduced the carrier jet fighter in Korea, it was clear that the swept wing was the next step in the development process if speeds of 690 mph were to be attained. Of course, sweeping the F9F-5's wing was a straightforward, simple problem in geometry and shop tooling. Well, not exactly; the aerodynamic effects of such a mechanical issue were significant since sweeping an otherwise flawless straight wing had well-known negative effects on maximum lift coefficient, aileron effectiveness, and flap effectiveness. It was looking to be an uphill battle for the Grumman company. Therefore, due to the F9F-6 conversion requiring extensive redesign, Grumman considered proposing an entirely new fighter for the Navy but felt it was in danger of losing out to North American and McDonnell, both of which had advanced swept-wing naval fighters. Subsequently, Grumman stuck to the Cougar.

On March 2, 1951, authorization was granted for the immediate construction of a prototype, and an order was placed for three experimental aircraft, designated XF9F-6. These prototypes retained the fuselage and tail assembly of the F9F series aircraft, but the wing was designed with a 35° sweepback and a 40% increase in wing area.

The first US Navy Grumman XF9F-6 Cougar (BuNo 126670) in flight, in 1951. This was early in the flight test program for this type. The prototype XF9F-6 exhibited superior carrier handling qualities than the straight-winged F9F-5. (US Navy)

The first XF9F-6 used a modified F9F-5 Panther fuselage and was powered by a 6,250 lb. thrust Pratt and Whitney J48-P-6A turbojet engine. It flew for the first time on September 20, 1951. The flight was conducted at Grumman's renowned Bethpage, New York facility, with Fred Rowley as the pilot. The XF9F-6 represented a significant improvement over the F9F-5. Its performance was instantly apparent, as its speed was around 85 mph faster than the F9F-5. However, the XF9F-6 was still a provisional design, and much work was required before the Navy could accept it.

F9F-6

F9F-6 Cougar (BuNo 128249) from Fighter Squadron VF-73 'Jesters'. VF-73 was assigned to Carrier Air Group 7 (CVG-7) aboard the aircraft carrier USS Hornet (CVA-12) during a deployment to the Western Pacific from May 4 to December 10, 1955. Squadron VF-73 was disestablished in 1958. (US Navy Photo, Collection of National Naval Aviation Museum)

The good news was that the fuselage of the Cougar was of excellent aerodynamic shape with a circular cross-section, similar to the Panther's fuselage. Moreover, the power plant arrangement was identical to that of the Panther. Still, the J48-P-4 or -6A turbojet of 6,250 lb. thrust (7,000 lb. with water injection) installed in the F9F-5 was replaced by the J48-P-8 of 7,250 lb. thrust (and 8,500 lb. with water injection) in all F9F-6 Cougar fighters, but not the early production models.

 No plane gets an easy ride in its initial stages of production, and the Cougar was no exception. Grumman created the F9F-6 Cougar by transforming the traditional straight wing of the F9F-5 Panther into a swept wing design, ushering in the transonic swept wing era. However, as mentioned earlier, the Cougar's swept wing, mid-placed horizontal tail, and thick wing section, inherited from the subsonic Panther, were ill-matched. They caused many issues for Grumman. Before the new Cougar could be put into fleetwide service, its wing had to be redesigned entirely due to the aircraft's serious pitch-up and departure characteristics at low and high speeds. Even after that, it continued to have several disagreeable traits, most notably a limited angle-of-attack range during carrier landing approaches that provided the pilot with a small maneuver window before the Cougar became unstable. Aware of the probable results of stalling and pitching up in the final seconds of flight before 'catching the wire' on an aircraft carrier, pilots chose to fly in faster. However, the increased safety came at the expense of less precise approaches with a higher risk of 'wave-offs' (aborted landings) and 'bolters' (touching down beyond the cables and having to accelerate back into the air, or worse, crashing into the parked aircraft).

Eventually, Grumman corrected the F9F-6's excessive landing speed by increasing the chord of the leading edge slats and the trailing edge flaps. These improvements were also implemented in the Panther. When the F9F-6s joined Squadron VF-32 'Swordsmen' in November 1952, they became the first swept-wing fighters in Naval service. Moreover, the F9F-6 Cougars and the following variants were the first military aircraft to have tubeless undercarriage tires. Being the early models, they were powered by a Pratt & Whitney J48-P-8 with water injection, 6,250lb st., 7,250lb ab.

The F9F-6's production was limited, although the -6 was available and in use by the conclusion of the Korean War. The Marine Corps received the majority of the early Cougars, and in addition to operating as fighters, a few were converted into F9F-6P reconnaissance aircraft. By 1958, the remaining F9F-6s had been delivered to reserve flying units. The total number of -6s produced was 646, with production ceasing on July 2, 1954.

Four US Naval Air Reserve Grumman F9F-6 Cougars of the Naval Air Reserve Training Unit (NARTU) Denver, flying over Pikes Peak, Colorado, on October 15, 1955. (US Navy Photo, Collection of National Naval Aviation Museum).

This F9F-6, BuNo 130929, was with squadron VF-191 'Satan's Kittens'. VF-191 was disestablished in 1978. The Cougar is seen here at San Francisco in August 1954. (Bill Larkins)

F9F-6, BuNo 128291, of the Glenview Naval Air Reserve Training Unit in the snow. (US Navy Photo, Collection of National Naval Aviation Museum)

An F9F-6 of fighter squadron VF-73 'Jesters'. VF-73 was assigned to Carrier Air Group 6 (CVG-6) aboard the aircraft carrier USS Midway (CVA-41) for a deployment to the Mediterranean Sea from January 8 to August 4, 1954. Squadron VF-73 was disestablished in 1958. (USN photo by John J. McFarland)

Two F9F-6 Cougars of squadron VF-21 'Freelancers' pictured during flight operations on board USS Randolph (CVA-15) in the Mediterranean Sea, in 1955. USS Randolph was decommissioned on February 13, 1969. (US Navy Photo, Collection of National Naval Aviation Museum).

F9F-6 Specifications

TYPE	POWERPLANT	DIMENSIONS WINGSPAN / LENGTH / HEIGHT	WEIGHTS Empty / Combat / Gross / Max. t.o.w.	SPEEDS Cruise / Max Sea Level / At 35,000 ft / Stalling Speed	SERVICE CEILING / RANGE
F9F-6	Pratt & Whitney J48-P-8 with water injection, 6,250 lb. s.t. (2,835kg), 7,250 lb. ab. (3,288kg)	34ft 6in (10.51m)/ 41ft 5in (12.62m)/ 12ft 4in (3.75m)	11,255lb. (5,105kg)/ 16,244lb. (7,368kg)/ 18,450lb. (8,369kg) / 21,000lb. (9,526kg)	541mph (871 kph)/ 654mph (1,053kph)/ 591 mph (951 kph) / 128mph (206kph)	44,500ft (13,564m) / 810mi (1,304km)
F9F-6 WEAPONS	Four 20mm cannon and 2,000 lb. ordnance on two wing pylons				

F9F-7

A Cougar variant was made available by Grumman, which was powered by the Allison J33-A-16A turbojet engine with a thrust of 6,350 lb. (or 7,000 lb. with water injection). This Cougar, built with the Allison engine as an alternate source of power plant supply for the Cougar, was designated F9F-7. There were no visible outward differences between the F9F-6 and F9F-7 Cougar versions due to the similar installation dimensions of the two engines.

The two Cougars performed similarly, with a peak speed of around 650 mph. In contrast to the F9F-6, this second Cougar model was only constructed as a day fighter. Although the cockpit arrangement was similar, it was instrumented to operate solely as a day fighter. The leading edge flaps were hydraulically linked to the flaps to function in tandem for enhanced handling qualities, particularly during landing. Most F9F-7s were assigned to Navy squadrons, and many were used by reserve units later. The former midnight blue color scheme for US Navy aircraft was changed to the new seagull grey and white scheme during the front line service days of the F9F-6s and -7s.

The first F9F-7 took to the air in March 1953, and the first delivery was made on April 15, 1953. A total of 168 F9F-7s were produced. The last F9F-7 was delivered on June 28, 1954.

F9F-7 SPECIFICATIONS

TYPE	POWERPLANT	DIMENSIONS WINGSPAN / LENGTH / HEIGHT	WEIGHTS Empty / Combat / Gross / Max. t.o.w.	SPEEDS Cruise / Max Sea Level / At 35,000 ft / Stalling Speed	SERVICE CEILING / RANGE
F9F-7	Allison J33-A-16A, with water injection, 6,350lb s.t. (2,880kg), 7,000lb ab. (3,175kg)	34ft 6in (10.51m)/ 41ft 5in (12.62m)/ 12ft 4in (3.75m)	11,483lb. (5,208kg)/ 16,244lb (7,368kg)/ 18,905lb (8,575kg)/ 21,000lb. (9,526kg)	509mph (819kph)/ 628mph (1,011kph)/ 559mph (900kph) / 130mph (209kph)	40,200ft (12,253m) / 1,157 mi (1,862km)
F9F-7 WEAPONS	Four 20mm cannon and 2,000 lb. bombs				

An F9F-7, BuNo 130848, at an unknown location. (Photo by Grumman via National Naval Aviation Museum).

F9F-7, BuNo 130777, assigned to the Naval Air Technical Training Unit (NATTU), seen here at Naval Air Facility (NAF) Mustin Field, in 1964. (US Navy Photo, Collection of National Naval Aviation Museum)

F9F-7, BuNo 130797, of Fighter Squadron VF-21 'Freelancers', in flight near Naval Air Station (NAS) Oceana, Virginia, on October 30, 1953. This F9F-7 was put into storage in February 1957, at NAF Litchfield Park, Arizona. (US Navy Photo, Collection of National Naval Aviation Museum)

F9F-7, BuNo 130761. This Cougar joined squadron VF-61 'Jolly Rogers' in 1953. This squadron was disestablished in 1959. This F9F-7 also served with the following squadrons: FASRON-3, FAGU El Centro, FASRON-12, and Navy Naval Air Reserve Training Unit (NARTU) New York. It was put into storage at NAF Litchfield Park in November 1956. (Grumman photo via National Naval Aviation Museum)

This is one of 168 F9F-7 Cougars manufactured by Grumman, BuNo 130776, seen here on June 10, 1954. From 1954-1956, it was with the Navy Naval Air Reserve Training Unit (NARTU) at New York as 7R-105. On May 30, 1960, it was officially removed from the register at NAS Norfolk. (US Navy Photo, Collection of National Naval Aviation Museum)

In 1956 this F9F-7 Cougar (BuNo 141043) performed refueling equipment tests with an A-3D Skywarrior. The Skywarrior was configured as a tanker at the Douglas test facility at Edwards AFB in California. During one 'plug' (refueling maneuver) with the A-3D, a section of the tubing detached from the Skywarrior but remained connected to the Grumman's front section. The F9F pilot was furious having to land on the lakebed with several feet of tubing hanging from the nose of his Cougar. The nerve of those Whale pilots. (US Navy)

F9F-6P

In 1952, Grumman introduced the F9F-6P photograph reconnaissance variant. The Navy bought a total of 60 F9F-6Ps after its great success with the F9F-2P and -5P photo reconnaissance aircraft. The F9F-6Ps shared the same camera configuration as the earlier Panthers deployed in Korea. There were seven cameras in all, and they could be used anytime, day or night.

The fuselage of the F9F-6P went from 41 feet 5 inches to 42 feet 17.8 inches in length with the new front segment. This front area was where the trimetrogon and K-17 cameras were situated. Both the Navy and the Marines flew the F9F-6Ps, which frequently had more colorful paint jobs than their all-blue fighter counterparts. The F9F-6P retained the same engine, fuel characteristics, and dimensions as the -6, but it weighed 250 pounds less than the fighter version. In addition, all weapons on the F9F-6P variant were removed.

The Navy used the Photo Reconnaissance Cougars from 1953 to the end of the 1950s. They were all painted in the dark sea blue scheme in the early years, but in 1957, some F9F-6P Cougars changed to the new grey and white scheme.

F9F-6P, BuNo 134458, of the Photographic Reconnaissance squadron VC-61, over NAS Miramar, California on March 4, 1954. (US Navy Photo, Collection of National Naval Aviation Museum)

A sailor is taking out a camera from the front of an F9F-6P of the Photo Reconnaissance squadron. Note the elongated front section of the aircraft. (US Navy Photo, Collection of National Naval Aviation Museum)

An F9F-6P, BuNo 127484, of the Marines, assigned to Marine Photographic Squadron Two (VMJ-2). VMJ-2 was founded on September 15, 1952, at MCAS Cherry Point, North Carolina. The McDonnell F2H-2P Photo Banshee was the squadron's workhorse, until 1953 when the F9F-6P Cougars arrived on the scene. The Cougars basically replaced almost all the Banshees by 1955. (National Archives)

F9F-6D/F9F-6K

After they were decommissioned from active service, several F9F-6 Cougars were used as target drones. They were given the designation F9F-6D, while others served as drone directors for combat training and were given the designation F9F-6K.

F9F-6K2 (QF-9G) this Panther, BuNo 128249, was an F9F-6, when it joined the navy in 1954. In 1961, it was converted to an F9F-6K2 target drone, and was based at China Lake, California. This photo was taken at NAS Norfolk on August 24, 1961. (US Navy Photo, Collection of National Naval Aviation Museum)

F9F-8/F8B

Following the Panther's success, Grumman launched the swept-wing F9F-6 Cougar, which became operational in November 1952. Then, in December 1953, just over a year later, the last version of the Cougar, the definitive F9F-8 model (Model G-99), was introduced. It differed from previous versions in that the fuselage was 25 cm longer, increasing the fuel tank capacity, which allowed more fuel to be stowed in each wing. Also, to improve rearward visibility, the canopy was modified. Regarding the wing, its chord (the distance between leading/front and trailing/back ends) was extended 15%, reducing its relative thickness and creating less drag. The chord was increased by way of a cambered leading edge (the measure of the curve of an airfoil from the leading edge to the trailing edge of the wing), replacing the moveable wing slats, and at the trailing edge, where the surface of the flaps were extended. It resulted in an eight-inch lengthening of the fuselage. This new design improved stability and handling at low speeds and high angles of attack, which was critical for aircraft carrier approaches and landings. Furthermore, this significant wing extension increased the F9F-8's internal fuel capacity to 1,063 gallons. The -8 could also carry two 150-gallon fuel tanks under the wings bringing the total fuel capacity up to 1,363 gallons.

The prototype F9F-8 was actually a converted F9F-6, just like the XF9F-6 was a modified F9F-5 Panther. Less than five months after the prototype's first flight, Grumman was able to launch the first production, F9F-8s, thanks to this time-saving factor.

In its production version, the F9F-8 was distinguished by the search radar, whose dome significantly protruded beneath the nose cone. The larger Pratt & Whitney J48-P-8A engine of 7,250 lb. thrust-8,800 lb. with water injection.

The Cougar -8 armament was similar to previous F9Fs, four fixed forward-firing M-3 20mm cannon, four AIM-9B Sidewinder AAMs, or four 500lb. bombs under the wings.

The F9F-8B version was basically a modified F9F-8, with cabling and cockpit equipment allowing it to launch either the early ASM-N-7 Bullpup short-range air-to-ground missiles or the AAM-N-7 Sidewinder air-to-air missiles, which were carried on four wing pylons outboard of the drop tanks. However, the US Navy made the modifications, not Grumman. These single-seat Cougars were used for a long time in secondary roles, such as the toss bombing tests for tactical nuclear bombs, officially known as the 'Low Altitude Bombing System' (LABS). In fact, the vast majority of F9F-8s were converted into F9F-8Bs capable of carrying and launching nuclear weapons. Overall, a total of 602 F9F-8s were produced by Grumman.

This F8F-8 Cougar (BuNo 141140) was upgraded to an F9F-8B. It is armed with what seems to be early versions of the Mark 82 227 kg bombs. (US Navy Photo, Collection of National Naval Aviation Museum)

F9F-8 Cougar, BuNo 14428, of VA-44 catches the wire on board the carrier USS Saratoga (CVA 60) on September 28, 1956. Converted to a QF-9J drone. In 1965 it joined VC-8 squadron 'Redtails'. In 1969 the Cougar was sent to its last posting at NAS China Lake, California. VC-8 was the Navy's last operator of the A-4 Skyhawk. The squadron was disestablished in 2003. (US Navy Photo, Collection of National Naval Aviation Museum)

An F9F-8 Cougar (BuNo 141140) armed with AIM-9B Sidewinder air-to-air missiles in 1958 (US Navy Photo, Collection of National Naval Aviation Museum)

This is F9F-8, BuNo 131120, of Fighter squadron 'VF-121' 'Peacemakers'. In 1967 it was with Training Squadron 'VT-21' 'Redhawks' as 3E-115. At the time of writing,'VT-21' was an active squadron. Unfortunately, this Cougar was damaged beyond repair after it suffered an accident on October 18, 1968. (US Navy Photo, Collection of National Naval Aviation Museum)

F9F-8 Cougar, BuNo 141124, of Fighter squadron VF-61 'Jolly Rogers'. Assigned to Carrier Air Group 8 (CVG-8) aboard the carrier USS Intrepid (CVA-11) for a deployment to the Mediterranean Sea from March 12 to September 15 1956. In 1956 the Cougar joined Attack Squadron VA-76 'Spirits'. VA-76 was disestablished in 1969. (Official US Navy photo)

T. Gianna. F9F-8, BuNo 141140, was upgraded to an F9F-8B ('toss' bomber). On June 27, 1970, it was struck off charge, at Naval Air Rework Facility (NARF) North Island. (T. Gianna, US Navy photo, provided by Ronnie Bell)

F9F-8 SPECIFICATIONS

TYPE	POWERPLANT	DIMENSIONS WINGSPAN / LENGTH / HEIGHT	WEIGHTS Empty / Combat / Gross / Max. t.o.w.	SPEEDS Cruise / Max Sea Level / At 35,000 ft / Stalling Speed	SERVICE CEILING / RANGE
F9F-8	Pratt & Whitney J48-P-8A with water injection. 7,250 lb s.t. (3,288kg). 8,500 lb w.i. (3,855kg).	34ft 6in (10.51m)/ 42ft 19/16in (12.85m)/ 12ft 3in (3.73m)	11,866lb (5,382kg)/ 17,328lb (7,860kg)/ 18,905lb (8,575kg)/ 24,763lb (11,232kg)	516mph (830kph)/ 642mph (1,033kph)/ 593mph (954kph) / 132mph (212kph)	42,000ft (12,802m)/ 1,312mi (2,111km)
F9F-8 WEAPONS	Four 20mm cannon and 2,000 lb. bombs, or four missiles				

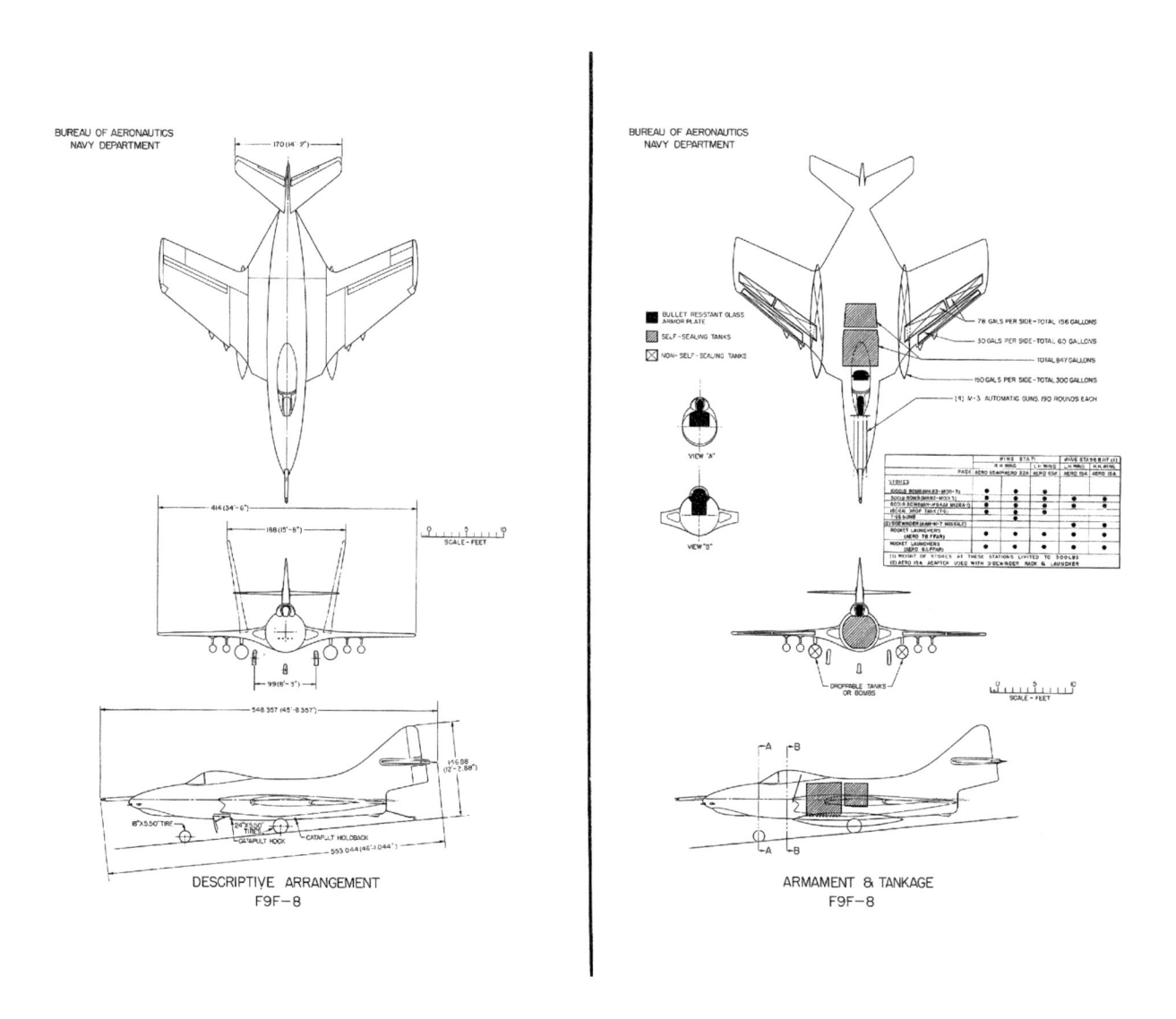

GRUMMAN F9F-8 COUGAR - STANDARD AIRCRAFT CHARACTERISTICS (Graphics: US Navy)

F9F-8/F8P

The Grumman Aircraft Engineering Corporation needed eight months and six days to develop the F9F-8P Photo Cougar, from the first mock-up to the prototype's first flight on August 21, 1955. To save time, and study the aerodynamic problems, while the prototype was being built, a normal F9F-8 was modified to 'F9F-8P' standard and flight tested. Grumman boasted that this F9F-8P could fly nonstop coast-to-coast across the United States and produce a continuous photograph of the entire route, spanning nearly ten miles.

The converted photo Cougar had a standard F9F-8 airframe, a lengthened fuselage, and an unusually shaped nose section. The nose of the Cougar housed 14 semi-automatic cameras manufactured by Fairchild Camera, Chicago Aerial Industries, and Instrument Corporation. Lockheed Martin acquired this company in 1996. The installation of photographic equipment necessitated the removal of the standard armament entirely. The fuel tanks were similar to the fighter version, with two 150-gallon fuel tanks located beneath the wings.

The -P version was powered by one Pratt and Whitney J48-P8A Turbo-Wasp engine of 7,250 lb. static thrust at take-off; with afterburner. The fact that the-P variant was devoid of weapons made it lighter and faster. Its speed was comparable to a standard F9F-8, about 712 mph. In total, 110 F9F-8Ps were manufactured. Deliveries began on August 29, 1955, and ended on July 12, 1957.

F9F-8P, BuNo 141668. This aircraft was the first of 110 F9F-8Ps. Its first flight was on February 18, 1955. (Grumman photo via National Naval Aviation Museum)

F9F-8T

This is YF9F-8T (YTF-9J), Buno 141667, model G-105. It was redesignated TF-9J in 1962. This was the only F9F-8 produced as a two-seat trainer prototype, armed with two 20mm cannon. It made its inaugural flight on February 29, 1956. Quite a rare photo. (US Navy Photo, Collection of National Naval Aviation Museum).

Almost 400 Cougars (399, to be exact) were manufactured as F9F-8T trainers. Grumman installed a new forward cockpit to create the trainer and removed two of the four 20 mm cannon and accessories. The cockpits were identical to those of the F9F-8 fighter and were built to the same standards. They were also equipped with two Martin-Baker ejector seats. Furthermore, they had the same water-injected Pratt & Whitney J48-P-8A engine as the F9F-8. Despite the increased total weight, the aircraft's performance only decreased slightly. The fuselage of the F9F-8T was extended by 23 inches, but no other dimensions were altered. The F9F-8's service ceiling of 50,000 feet was reduced to 40,000 for the trainer, and the combat radius was whittled down to 280 miles. The good news was that the top speed and range remained virtually unchanged. Even though the armament was reduced to two 20 mm cannon, up to 2,000 pounds of external stores could still be carried. The F9F-8T trainer version of the F9F-8 Cougar took to the skies for the first time on April 4, 1956.

Between 1956 and 1960, the Navy purchased 377 of the F9F-8Ts. They were used for carrier and advanced flight training, as well as weapons training, given that they were armed with two 20mm cannon and could carry bombs or missiles.

In 1962, the -8T was renamed the TF-9J. In February 1974, Training Squadron VT-4 'Warbucks' was one of the last units to fly the TF-9J. Yet, oddly, it was the F9F-8T training version that flew the Cougar into hostile skies, with a small number of them serving as Forward Air Control (FAC) aircraft with Marine Headquarters and Maintenance Squadron (H&MS) 11 during the Vietnam conflict.

TF-9J Cougars of Headquarters and Maintenance Squadron 11 (H&MS-11) at Da Nang Air Base, Vietnam. H&MS-11 and H&MS-13 used the Cougar as a fast Forward Air Control (FAC) aircraft until being replaced by the Douglas TA-4F Skyhawk. Only during the Vietnam War did the Cougars engage in combat. The Cougar was in service with the Navy from 1952 to 1974. (From the John T. Dyer Collection (COLL/3503) at the Archives Branch, Marine Corps History Division.)

A plane captain directs an F9F-8T Cougar of Marine squadron VMT-2 at MCAS El Toro, California, #OTD in 1959. (US Navy Photo, Collection of National Naval Aviation Museum)

A TF-9J Cougar advanced trainer assigned to Training Squadron 24 (TRARON 24) VT-24 'Bobcats', pictured during a cross-country flight to Scott AFB, Illinois, #OTD in 1971. VT 24 squadron was disestablished in 1992. (US Navy Photo, Collection of National Naval Aviation Museum)

Two US Marine Corps Grumman F9F-8T Cougars (BuNo 147317, 147328) from Marine Training Squadron VMT-1, 2nd Marine Aircraft Wing, in flight near Marine Corps Air Station Cherry Point, North Carolina (USA), in 1962. (US Navy Photo, Collection of National Naval Aviation Museum)

Two TF-9Js of VF-126 squadron 'Bandits' ca.1965. The top aircraft, BuNo 147382, NJ620, was with Fleet Marine Force, Pacific (AIR FMF PAC) as WZ-2, in the early 1960s. The other aircraft, BuNo 147376, was with VF-126 as NJ-624. Then, from 1969 to 1970 it flew with VT-22 squadron 'Golden Eagles' as 3F-279. It was based at Naval Auxiliary Air Station (NAAS) Kingsville, Texas. It was finally put into storage at the AMARC bone yard in July 1974. VF-126 was disestablished on April 1 1994. VT 22 is still an active squadron. (US Navy Photo, Collection of National Naval Aviation Museum)

Two Cougars of VF-174 squadron 'Hell Razors' getting ready for takeoff. VF-174 was redesignated VA-174 in 1966. (US Navy Photo, Collection of National Naval Aviation Museum)

F9F-8T/TF-9J SPECIFICATIONS

TYPE	POWERPLANT	DIMENSIONS WINGSPAN / LENGTH / HEIGHT	WEIGHTS Empty / Combat / Gross / Max. t.o.w.	SPEEDS Cruise / Max Sea Level / At 35,000 ft / Stalling Speed	SERVICE CEILING / RANGE
F9F-8T/TF-9J	Pratt & Whitney J48-P-8A with water injection. 7,200lb s.t. (3,266kg). 8,500lb w.i. (3,855kg).	34ft 6in (10.54m)/ 48ft.65in (14.82m)/ 12ft 1in (3.68m)	12,768lb (5,791kg)/ 17,790lb (8,069kg) / (20,600lb (9,344kg)/ 24,178lb (10,966kg)	630mph (1,014kph)/ 435mph (700kph)/ 482mph (775kph) / 124mph (200kph)	40,400ft (12,313m) / 600mi (966km)
F9F-8T/TF-9J WEAPONS	Two 20mm M-3 cannon				

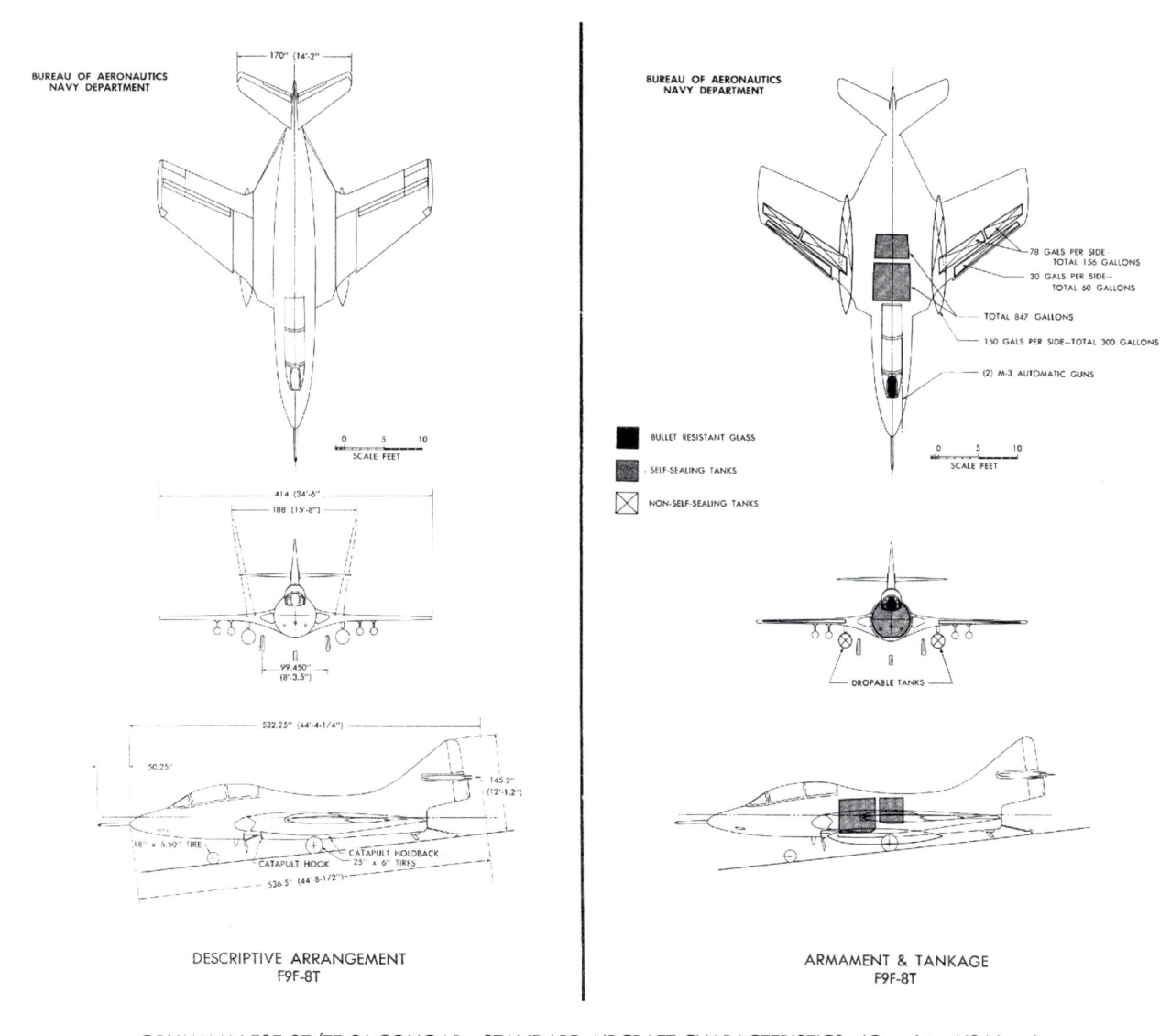

GRUMMAN F9F-8T/TF-9J COUGAR - STANDARD AIRCRAFT CHARACTERISTICS   (Graphics: US Navy)

67

CHINA LAKE

The Naval Air Weapons Station (NAWS) China Lake is a vast military site in California that serves the United States Navy's research, testing, and evaluation activities. It is part of Navy Region Southwest and is directed by the Commander, Navy Installations Command.

Back in the 1950s, it was the Navy's biggest ordnance research and development facility. Work at the permanent field station of the Bureau of Ordnance provided rockets, guided missiles, torpedoes, and aircraft fire-control systems to the Navy and other combat forces in the United States. Civilian scientists and engineers generated new weapon designs and saw them through the development cycle until they were ready for mass manufacturing.

Military personnel offered operational know-how and brought the demands of the Fleet to the Station's attention. This civilian-military team was concerned not just with urgent needs but also with weapon systems that would be needed in the future.

China Lake was also where naval aircraft near retirement were sent to be used as drones. Many Panthers and Cougars ended up at China Lake. It was a gloomy way to end their service life.

Grumman QF-9J Cougar (BuNo 138886) of the 'Redbird' Squadron at Naval Air Weapons Station, China Lake, California, rests on a flight line. After a lengthy and diverse service life, this aircraft was part of the NOLO (No Live Operator) program. It completed twenty successful NOLO flights before crashing on the twenty-first flight. In its early service life, it was upgraded to an F9F-8B, and was with VT-21 squadron 'Redhawks'. (US Navy Photo, Collection of National Naval Aviation Museum)

F9F-6 Cougars of VF-24 squadron 'Corsairs', operate on board USS Yorktown (CVA 10) during operations off Southern California in March 1954.

The Navy received 1,388 Panthers with the designations F9F-2, F9F-3, F9F-4, and F9F-5. The Panther quickly became a Navy icon after being the first Naval fighter to shoot down an enemy aircraft during the Korean War. Not bad for the first Grumman jet to see service with the US Navy. Following the Panther's success, Grumman created a swept wing version, and proposed it to the Navy. The Navy eagerly adopted the 'Cougar' model. After all, Grumman claimed the Cougar was more potent than the Panther.

The Cougar was initially delivered to the Navy in November 1952 and continued in navy squadron service until February 1960. A total of 1,985 Cougars were manufactured between the original XF9F-6 and the last F9F-8 trainer. Only the Grumman Hellcat, Corsair, and Wildcat fighters manufactured during WWII surpassed this outstanding postwar total. The total number of Panthers and Cougars in US Navy service reached 3,373.

F9F Panthers of ATU-206 (Advanced Training Units), at NAS Pensacola in 1956. (US Navy Photo by Joseph A. Gryson; Collection of National Naval Aviation Museum)

A US Navy Grumman F9F-2 Panther (BuNo 122567) in flight. This was the eighth production aircraft. It was with NATC, at NAS Patuxent River in Maryland. (US Navy Photo, Collection of National Naval Aviation Museum)

This is an F9F-2 First production version, powered by the Pratt & Whitney J42 engine. Seen here at the Marine Corps Air Station Miramar (MCAS Miramar) in 2014. (Photo Tomás Del Coro)

Grumman F9F-8P, (RF-9J) (BuNo 144424), assigned to Fighter Photographic Squadron VFP-61 clearly shows the modified nose section of the photo reconnaissance version of the Cougar. Previously, in 1959, this aircraft was with Composite Photographic Squadron Sixty-Three VCP-63 as PP-964. It was put into storage at NAF Litchfield Park, before being struck off the charge on October 16, 1963. (US Navy Photo, Collection of National Naval Aviation Museum)

Grumman F9F Panther being inspected by the Midshipmen at Naval Air Test Center, Patuxent River, Maryland, circa 1950s. (US Navy Photo, Courtesy of Commander Richard Timm, Naval History and Heritage Command)

Looking at the pack on the bow of USS Coral Sea CVA-43 with F9F-8B Cougars. (photo from 'usscoralsea.net' by Doug Riach of the US Navy on the USS Coral Sea in 1956.)

Grumman TF-9J Cougar (BuNo 146405) of Training Squadron VT-10 'Wildcats' at Naval Air Station Miramar, California (USA), on October 13, 1973. VT-10 was the last Cougar squadron to fly the TF-9J Cougar until 1974. This Cougar may well have been the one to make the final flight of the Cougar, ending 22 years of service with the Navy for the enigmatic Panthers and Cougars. (US Navy Photo, Collection of National Naval Aviation Museum)

An F9F Panther roars past the island of the carrier Oriskany (CVA-34) while making touch and go landings on July 27, 1953. (US Navy Photo, Collection of National Naval Aviation Museum)

F9F-2 Panther (BuNo123030) of Fighter Squadron 24 (VF-24) 'Corsairs' over Koo Island, South Korea, on June 27, 1952. Assigned to USS Boxer (CV-21), Pilot Lt (JG) G.W. Stinnett, Jr. Squadron VF-24 was redesignated VF-211 in 1959. (National Archives)

KOREAN WAR F9F SQUADRONS

DEPARTURE DATE	RETURN DATE	AIR WING	SQUADRON	AIRCRAFT TYPE	TAIL CODE	AIRCRAFT CARRIER
May 12, 1951	Aug, 29, 1951	CVG-19	VF-23	F9F-2	B	CV-37 USS Princeton
June 16, 1952	Feb. 6 1953	ATG-2	VF-23	F9F-2	M	CV-37 USS Princeton
Feb. 8, 1952	Sep. 26, 1952	CVG-2	VF-24	F9F-2	M	CV-21 USS Boxer
Sep 6, 1950	Feb. 3, 1951	CVG-3	VF-31	F9F-2	K	CV-32 USS Leyte
June 26, 1951	March 25, 1952	CVG-5	VF-51	F9F-2	S	CV-9 USS Essex
March 30, 1953	Nov. 28, 1953	ATG-1	VF-52	F9F-2	S	CV-9 USS Essex
Nov. 20, 1952	June 25, 1953	CVG-5	VF-51	F9F-5	S	CV-45 USS Valley Forge
Oct. 15, 1951	July 3, 1952	ATG-1	VF-52	F9F-2B	S	CVA-45, USS Valley Forge
May 1, 1950	Dec. 1, 1950	CVG-5	VF-51	F9F-3	S	CV-45 USS Valley Forge
May 1, 1950	Dec. 1, 1950	CVG-5	VF-52	F9F-3	S	CV-45 USS Valley Forge
March 30, 1953	Nov. 28, 1953	ATG-1	VF-52	F9F-2	S	CVA-21 USS Boxer
Nov. 20, 1952	June 25, 1953	CVG-5	VF-53	F9F-5	S	CVA-45 USS Valley Forge
May 20, 1952	Jan. 8, 1953	CVG-7	VF-71	F9F-2	L	CV-31 USS Bon Homme Richard
May 20, 1952	Jan. 8, 1953	CVG-7	VF-72	F9F-2	L	CV-31 USS Bon Homme Richard
Dec. 15, 1952	Aug. 14, 1953	CVG-9	VF-91	F9F-2	N	CVA-47 USS Philippine Sea
Dec. 15, 1952	Aug. 14, 1953	CVG-9	VF-93	F9F-2	N	CVA-47 USS Philippine Sea
July 5, 1950	April 7, 1951	CVG-11	VF-111	F9F-2B	V	CV-47 USS Philippine Sea
Oct. 15, 1951	July 3, 1952	ATG-1	VF-111	F9F-2B	V	CV-45 USS Valley Forge
April 26, 1953	Dec. 4, 1953	CVG-4	VF-111	F9F-5	V	CVA-39 USS Lake Champlain
March 30, 1953	June 30, 1953	ATG-1	VF-111	F9F-5	V	CVA-21 USS Boxer
Dec. 31, 1951	Aug. 8, 1952	CVG-11	VF-112	F9F-2	V	CV-47 USS Philippine Sea
July 5, 1950	March 26, 1951	CVG-11	VF-111	F9F-2	V	CV-47 USS Philippine Sea
July 5, 1950	March 26, 1951	CVG-11	VF-112	F9F-2	V	CV-47 USS Philippine Sea
March 30, 1953	Nov. 28, 1953	ATG-1	VF-151	F9F-2	H	CVA-21 USS Boxer
Jan. 24, 1953	Sep. 21, 1953	CVG-15	VF-153	F9F-5	H	CVA-37 USS Princeton
Jan. 24, 1953	Sep. 21, 1953	CVG-15	VF-154	F9F-5	H	CVA-37 USS Princeton
Nov. 9, 1950	May 29, 1951	CVG-19	VF-191	F9F-2	B	CVA-37 USS Princeton
Jan. 24, 1953	Sep. 21, 1953	CVG-15	VF-191	F9F-5	H	CV-37 USS Princeton
March 21, 1952	Nov. 3, 1952	CVG-19	VF-191	F9F-2	B	CV-37 USS Princeton
March 2, 1951	Oct. 24, 1951	CVG-101	VF-721	F9F-2B	A	CV-21 USS Boxer
Aug.11, 1952	March 17, 1953	CVG-101	VF-721	F9F-2	A	CV-33 USS Kearsarge
May 10, 1951	Dec. 17, 1951	CVG-102	VF-781	F9F-2	D	CV-31 USS Bon Homme Richard
Sep. 15, 1952	May 18, 1953	CVG-102	VF-781	F9F-2B	D	CVA-34 USS Oriskany
Sep. 15, 1952	May 18, 1953	CVG-102	VF-783	F9F-5	D	CVA-34 USS Oriskany
June 16, 1952	Feb. 6 1953	ATG-2	VF-821	F9F-2	A	CV-9 USS Essex
Sep. 8, 1951	May 2, 1952	CVG-15	VF-831	F9F-2B	H	CV-36 USS Antietam
Sep. 8, 1951	May 2, 1952	CVG-15	VF-837	F9F-2B	H	CV-36 USS Antietam

GRUMMAN F9F PRODUCTION LIST

MODEL	BUREAU NUMBER	UNITS PRODUCED
XF9F-2	122475	1
XF9F-3	122476	1
XF9F-2	122477	1
F9F-2	122614-122708	95
F9F-3	123016-123083	68
XF9F-4	123084	1
XF9F-5	123085	1
F9F-3	123068-123087 (123087 canceled)	19
F9F-2	123397-123740 (123714-123740 canceled)	318
F9F-5	125080-125152	73
F9F-4	125153-125225	73
F9F-5	125226-125313	88
F9F-5P	125314-125321	8
F9F-5	125414-125443	30
F9F-5	125489-125499	11
F9F-5	125533-125648	115
F9F-5	125893-126256	364
F9F-6	126257-126264	8
F9F-5P	126265-126290	26
F9F-5	126627-126669	43
XF9F-6	126670-126672	3
F9F-2	127086-127215	130
F9F-6	127216-127470	255
F9F-5P	127471-127472	2
F9F-6P	127473-127492	20
F9F-6	128055-128294	240
F9F-6P	128295-128310	16
F9F-7	130752-130919	168
F9F-6	130920-131062	143
F9F-8	131063-131251	189
F9F-6P	131252-131255	4
F9F-8	134234-134244	11
F9F-6	134245-134433	189
F9F-6P	134446-134465	20
F9F-8	138823-138898	76
F9F-8	141030-141229	200
F9F-8	141648-141666	19
YF9F-8T	141667	1
F9F-8P	141668-141727	60
F9F-8T	142437 142532	96
F9F-8T	142945-143013	69
F9F-8	144271-144376	106
F9F-8P	144377-144426	50
F9F-8T	146342-146425	84
F9F-8T	147270-147429	**160**

Other Image Credits

Page 9 **Andrew Thomas** (https://www.flickr.com/photos/atom-uk/5840641038/in/photolist-9U7Nvo-9PHQTn-9U4Z3c-9U7P6y-aWhYXg). https://creativecommons.org/licenses/by/2.0/legalcode

Page 30 **Bill Larkins** (https://www.flickr.com/photos/34076827@N00/1013003909/in/photolist-vqZYY5-2xvULD-biHWkM-biHWjX), https://creativecommons.org/licenses/by/2.0/legalcode (Colorized by Theodore Gianna)

Page 37 **RuthAS** (https://commons.wikimedia.org/wiki/File:Grumman_DF-9E_126277_VU-1_ONT_181070_edited-2.jpg), „Grumman DF-9E 126277 VU-1 ONT 181070 edited-2", sharper, more color brighter by Theodore Gianna, https://creativecommons.org/licenses/by/3.0/legalcode

Page 48 **Bill Larkins** (https://www.flickr.com/photos/34076827@N00/6764130507/in/photolist-vqZYY5-2xvULD-biHWkM-biHWjX), https://creativecommons.org/licenses/by/2.0/legalcode (Colorized by Theodore Gianna)

Page 64 **USMC Archives** (https://commons.wikimedia.org/wiki/File:TF-9J_Cougars_49716196041.jpg), „TF-9J Cougars 49716196041", Brighter and some saturation by Theodore Gianna, https://creativecommons.org/licenses/by/2.0/legalcode

Page 71 **Tomás Del Coro** from Las Vegas, Nevada, USA (https://commons.wikimedia.org/wiki/File:Grumman_F9F-2_Panther_123652_VMF-311_(14985328394).jpg), „Grumman F9F-2 Panther 123652 VMF-311 (14985328394)", https://creativecommons.org/licenses/by-sa/2.0/legalcode

Made in United States
Troutdale, OR
03/24/2025